Three-Dimensional Surface Topography; Measurement, Interpretation and Applications

A Survey and Bibliography

Three Dimensional Surface Topography; Measurement, Interpretation and Applications

A Survey and Bibliography

Editor: K J Stout
Centre for Metrology
School of Manufacturing and Mechanical Engineering
The University of Birmingham

Contributors

W P Dong, E Mainsah, L Blunt, K J Stout
The University of Birmingham, UK

N Luo
UBM Corporation, Roselle, NJ, USA

P J Sullivan
National Institute of Standards and Technology,
Gaithersburg, USA

Penton Press
London and Bristol, Pennsylvania

First published in the United Kingdom in 1994 by
Jessica Kingsley Publishers Ltd
116 Pentonville Road
London N1 9JB, England
and
1900 Frost Road, Suite 101
Bristol, PA 19007, U S A

Library of Congress Cataloging in Publication Data

A CIP catalogue record for this book is available
from the Library of Congress

British Library Cataloguing in Publication Data

A CIP catalogue record for this book is available
from the British Library on request

ISBN 1 85718 004 6

Printed and Bound in Great Britain by
Biddles Ltd., Guildford and King's Lynn

CONTENTS

Part I: Instruments and Measurement Techniques of 3-Dimensional Surface Topography 1

W P Dong, E Mainsah, P J Sullivan and K J Stout

1.1 Introduction 3

1.1.1 3-D, Areal, and Parametric Measurement 4 1.1.2 History of Surface Measurement 4

1.2 Differences in the Measurement and Analysis Methods for 2-D and 3-D Surface Topography 6

1.3 Stylus Instruments 10

1.3.1 Mechanism of 3-D Measurement 10 1.3.2 Construction of the Profile System 13 1.3.3 Stylus 3-D Systems 17

1.4 Optical Instruments 19

1.4.1 Focus Detection Instruments 19

1.4.1.1 Intensity Detection Method 21 1.4.1.2 Differential Detection Method 23 1.4.1.3 Critical Angle Method 23 1.4.1.4 Astigmatic Method 24 1.4.1.5 Foucault Method 25 1.4.1.6 Skew Beam Method 26 1.4.1.7 Defect-of-Focus Method 28 1.4.1.8 Confocal Method 28 1.4.1.9 Properties of Focus Detection Methods 30

1.4.2 Interferometric Instruments 32

1.4.2.1 Phase Shifting Interferometric Instrument 34 1.4.2.2 Scanning Differential Interferometric Instrument 37 1.4.2.3 Properties of Interferometric Instruments 39

1.5 Non-Optical Scanning Microscopy 41

1.5.1 Electron Microscopy 41

1.5.1.1 Stereopair Technique for Quantizing SEM Images 43 1.5.1.2 Direct Integration for Quantizing SEM images 45 1.5.1.3 Properties of Electron Microscopy 46

1.5.2 Scanning Probe Microscope 46

1.5.2.1 Scanning Tunnelling Microscope 48 1.5.2.2 Atomic Force Microscope 50 1.5.2.3 Properties of STM and AFM 53 1.5.2.4 Scanning Capacitance Microscopy 54

1.6 General Comments on the Different Types of Instruments 58

1.6.1 Measurement Range and Resolution 58

1.6.2 Measurement Speed 60

1.6.3 Problems 60

1.6.4 Application Areas 61

1.7. Conclusions 63

Part II: Three-Dimensional Surface Topography – Review of Present and Future Trends 65

W P Dong, E Mainsah, K J Stout and P J Sullivan

2.1 Introduction 67

2.2 Aims of the Survey 68

2.3 Main Findings of the Survey 69

2.3.1 3-D Versus 2-D Analysis 69   2.3.2 The Scope of Surface Topography 69   2.3.3 Instrumentation 70

2.3.3.1 Stylus-based Systems 70   2.3.3.2 Optical Systems 70 2.3.3.3 The Scanning Microscope 72

2.3.4 Digitisation – Range and Resolution 73

2.3.5 Levelling of Stages 73   2.3.6 Specimen Relocation 74

2.3.7 Measurement Datum Plane Definition 76   2.3.8 Static and Dynamic Measurement 76   2.3.9 Data Logging Conditions 77   2.3.10 Digital Filtering 77   2.3.11 Characterisation Reference Datum Plane 78

2.3.12 Characterisation and Parameters 78

2.3.12.1 Statistical Characterisation 79   2.3.12.2 Characterisation Via Spectral Analysis 79   2.3.12.3 Time Series Analysis 79 2.3.12.4 Functional Characterisation 80   2.3.12.5 Visual Inspection 83 2.3.12.6 Fractal Characterisation 83

2.3.13 Parameter Rash? 83

2.4 Conclusions 84

Part III: Visualisation Techniques and A Primary Parameter Set for Characterising Three-Dimensional Surface Topography

W P Dong, N L Luo, P J Sullivan and K J Stout

Nomenclature 89

3.1 Introduction 90

3.2 Surface Topography in Three-Dimensions 91

3.3 Reference Datum for Topographic Analysis 91

3.4 Visualisation Techniques 92

3.4.1 Visualisation Plots 93

3.4.1.1 Isometric Plot 93   3.4.1.2 Contour Plot 95   3.4.1.3 Greyscale Image 97

3.4.2 Manipulation Techniques 98

3.4.2.1 Inversion 98   3.4.2.2 Truncation 98 3.4.2.3 Zooming and Clipping 100 3.4.2.4 Surface Image Enhancement 101

3.5 Specification of the Parameters 103

3.6 A Primary Parameter Set 103

3.6.1 Amplitude and Height Distribution Parameters 104

3.6.1.1 Root-Mean-Square Deviation S_q 104   3.6.1.2 Ten Point Height of the Surface S_z 104   3.6.1.3 Skewness of Surface Height Distribution S_{sk} 106 3.6.1.4 Kurtosis of Surface Height Distribution S_{ku} 106

3.6.2 Spatial Parameters 106

3.6.2.1 Density of Summits of the Surface S_{ds} 107   3.6.2.2 Texture Aspect Ratio of the Surface S_{tr} 108   3.6.2.3 Texture Direction of the Surface S_{td} 113 3.6.2.4 Fastest decay autocorrelation length S_{al} 114

3.6.3 Hybrid Parameter 115

3.6.3.1 Root-Mean-Square Slope of the Surface $S_{\Delta q}$ 115   3.6.3.2 Arithmetic Mean Summit Curvature of the Surface S_{sc} 115   3.6.3.3 Developed Interfacial Area Ratio S_{dr} 116

3.6.4 Functional Parameters for Characterising Bearing and Fluid Retention Properties 118

3.6.4.1 Surface Bearing Index S_{bi} 119   3.6.4.2 Core Fluid Retention Index S_{ci} 119   3.6.4.3 Valley Fluid Retention Index S_{vi} 119

3.7 Conclusions 121

Part IV: Applications of Three-Dimensional Surface Metrology 123

W P Dong, L Blunt, K J Stout and P J Sullivan

4.1 Introduction 125

4.2 Measurement of A Gear Surface with the Stylus Lead Screw Driven Instrument 128

4.2.1 Instrument and Topography Measurement of A Gear Surface 129

4.2.2 Characterisation of the Gear Surface 129

4.3 Measurement of An Engine Bore Surface with the Stylus Linear Motor Driven Instrumen 133

4.3.1 Instrument and Topography Measurement of An Engine Bore Surface 134

4.3.2 Characterisation of the Engine Bore Surface 135

4.4 Measurement of Thick Film Superconductors with the Focus Detection Instrument 137

4.4.1 Instrumentation 138

4.4.2 Fabrication of the Thick Film Superconductors 139

4.4.3 Topography of the Thick Film Superconductors 139

4.5 Measurement of Human Skin with A Focus Detection Instrument 143

4.5.1 Replication of Surfaces 144

4.5.2 Topography of the Skin Surface 144

4.6 Measurement of the Topography of Hip Prostheses Using Phase Shifting Interferometer 145

4.6.1 Instrument and Topography Measurement of Hip Prosthesis Surfaces 147

4.6.2 Characterisation of the Hip Joint Surfaces 147

4.7 Measurement of Polished Brass Surface Using a Scanning Tunnelling Microscope 149

4.7.1 Instrument and Surface Measurement of a Polished Brass Surface 150

4.7.2 Topography of the Polished Brass Surface 151

4.8 Characterisation of Surface Topography of Indentations 153

4.8.1 Visual Characterisation 154

4.8.2 Numerical Characterisation 160

4.9 Conclusions 163

Appendix 1a: Technical Specification of Some 3D Topography Instruments 165

1A.1: Stylus-based Contacting Systems 165

1A.1.1 3-D Automated Surface Topography Analysis System (ASTA) 165

1A.1.2. Perthometer S8P 167

1A.1.3. Surfcom 475/575-3D 169

1A.1.4. Surfascan 3-D 172

1A.1.5. Form Talysurf Series 175

1A.2: Optical and Other Non-contacting Systems 177

1A.2.1. Micromap 512 Optical Profiler 177

1A.2.2. Wyko Topo-3D 180

1A.2.3. Wyko RST 182

1A.2.4. UBM Optical Surface Measurement System 184

1A.2.5. Rodenstock RM-600 3-D 187

1A.2.6. Scanning Force Microscope (SFM) 189

1A.2.7. Maxim 3-D Model 5700 190

1A.2.8. MP2000 193

1A.2.9. Proscan 1000 195

Appendix 2a: Survey Analysis 199

Part V Bibliography 205

Part I 205

Part II 220

Part III 223

Part IV 226

Index 229

LIST OF FIGURES

Figure 1.1   2-D and 3-D plots of a ground surface   7
                     (a) Profile of the ground surface
                     (b) Isometric plot of a truncated ground surface
Figure 1.2   A grey scale map of a coin logged by a 3-D stylus system   9
Figure 1.3   Coordinates of scan modes   11
                     (a) Coordinates of raster scan
                     (b) Coordinates of radial scan
Figure 1.4   An isometric plot of an inhomogeneous surface logged
          using different sampling interval in X and Y directions   12
Figure 1.5   Plot of traverse velocity versus position   13
Figure 1.6   Schematic diagrams of 2-D and 3-D stylus systems   14–15
                     (a) A conventional 2-D system (from Dagnall)
                     (b) A 3-D system with one translation stage and
                          a gear box
                     (c) A 3-D system with two translation stages
Figure 1.7   Optical arrangements of focus detection systems   20
                     (a) A general focus detection system
                     (b) A confocal system
Figure 1.8   Schematic diagram of an intensity focus detection system   21
Figure 1.9   Schematic diagram of a differential focus detection system   22
Figure 1.10 Schematic diagram of a critical angle focus detection system   24
Figure 1.11 Schematic diagram of an astigmatic focus detection system   25
Figure 1.12 Schematic diagram of a Foucault focus detection system   26
Figure 1.13 Schematic diagram of a skew beam focus detection system   27
Figure 1.14 Schematic diagram of a defect-of-focus focus
          detection system   27
Figure 1.15 Schematic diagram of the Tandem Scanning Microscope
          (from McCormick)   29
Figure 1.16 An example of optical fringe   33
Figure 1.17 Schematic diagram of the Michelson interferometer   33
Figure 1.18 Schematic diagram of a phase shifting interferometry
          instrument (from Bhushan)   36
Figure 1.19 Schematic diagram of a scanning differential interferometry
          instrument   38
                     (a) Construction of the instrument (from Bristow)
                     (b) Profiling principle of the Nomarski prism
                          (From Lessor)

Figure 1.20   Schematic diagram of the SEM (from Johari)   42
Figure 1.21   The derivation of the standard parallax equation
               (from Hudson)   44
Figure 1.22   Block diagram of a direct integration SEM system
               (from Sato)   45
Figure 1.23   Principle of the STM (from Besenbacher)   47
Figure 1.24   Electron clouds on surfaces of the tip and the sample
               (from Burleigh Instruments)   48
Figure 1.25   Schematic diagram of a STM system (from Digital
               Instruments)   51
Figure 1.26   Schematic diagram of a AFM (from Binnig)   52
Figure 1.27   Schematic diagram of a capacitance probe (from Matey)   55
Figure 1.28.   Schematic diagram of a scanning capacitance microscope
               (from Bugg)   57
                        (a) Schematic diagram of the probe and measurement
                        (b) Block diagram of the main components
Figure 1.29   Estimated amplitude-wavelength plot of 3-D systems   59
Figure 2.1   Example of the resolution of stylus instrument   75
                        (a) relation between vertical resolution and ADC
                            number of bits
                        (b) relation between FRD and Vmag
Figure 3.1   Co-ordination of a digitised surface   91
Figure 3.2   Isometric plots of a face turned surface with different
               projection angles   94
                        (a) Projection andgle is 45°
                        (b) Projection angle is 70°
Figure 3.3   Isometric plots of a bored surface with different rotation
               angles   95
                        (a) Rotation angle is 0°
                        (b) Rotation angle is 90°
Figure 3.4   Contour plots of a shaped surface   96
                        (a) The number of contour levels is 4
                        (b) The number of contour levels is 20
Figure 3.5   Intensity plots of a honed surface   97
                        (a) The number of grey levels is 16
                        (b) The number of grey levels is 64
Figure 3.6   Inversion of a ground surface   98
Figure 3.7   Truncation of an EDM surface   99
                        (a) 40% truncation
                        (b) 60% truncation
Figure 3.8   Top 40% material display of the EDM surface shown
               in Figure 3.7   100
Figure 3.9   Illustration of the zoom technique   100–101
                        (a) Selection of the interested area
                        (b) Re-display the data of the interested area

Figure 3.10 Contour plot of a honed surface which is clipped in the
             dark area   101
Figure 3.11 Image enhancement   102
             (a) Height image
             (b) Slope image in x direction
             (c) Slope image in y direction
             (d) Slope image in both x & y directions
Figure 3.12 Diagrams of S_z versus S_q for typical engineering surfaces 105
Figure 3.13 Diagrams of S_{ku} versus S_{sk} for typical engineering surfaces 107
Figure 3.14 Isotropic topography of an EDM surface   109–110
             (a) Isometric plot
             (b) AACF
             (c) APSD
             (d) Angular spectrum
Figure 3.15 Linear texture of a shaped surface   110–111
             (a) Isometric plot
             (b) AACF
             (c) APSD
             (d) Angular spectrum, S_{td} = -21°
Figure 3.16 Crossed lay texture of a plateau honed surface   112–113
             (a) Isometric plot
             (b) AACF
             (c) APSD
             (d) Angular spectrum, S_{td} = 19° and S_{td} = -21°
Figure 3.17 Definition of the texture direction   113
Figure 3.18 Schematic diagram of the interfacial area   116
Figure 3.19 Diagrams of the developed area ratio S_{dr} versus S_q of
             typical engineering surfaces   117
Figure 3.20 A bearing area ratio curve scaled according to the RMS
             derivation   118
Figure 3.21 A diagram of the valley fluid retention index S_{vi} versus
             the surface bearing index S_{bi} of typical engineering
             surfaces   120
Figure 3.22 A diagram of the valley oil retention index S_{vi} versus the
             core fluid retention index S_{ci} of typical engineering surfaces 120
Figure 4.1   Photograph of a 3-D lead screw driven stylus instrument
             (modified Talysurf 6) in measuring a gear tooth   128
Figure 4.2   Topography of the gear tooth surface measured by a stylus
             instrument   129
             (a) Original topography
             (b) Decurved topography
             (c) Grey scale image
Figure 4.3   Bearing area ratio and void volume ratio of the gear tooth
             surface   131
             (a) Bearing area ratio
             (b) Void volume ratio

Figure 4.4   Surface height distribution of the gear tooth surface   132
Figure 4.5   Photograph of a 3-D linear motor driven stylus instrument
              (DSAGE-3D) in measuring an engine bore   134
Figure 4.6   Topography of the engine bore measured by a stylus
              instrument   135
                  (a) Decurved topography
                  (b) Grey scale image
Figure 4.7   Surface height distribution of the engine bore surface   136
Figure 4.8   Bearing area ratio and void volume ratio of the engine bore   137
                  (a) Bearing area ratio
                  (b) Void volume ratio
Figure 4.9   A 3-D focus detection instrument (Rodenstock 600) in
              measuring superconducting material   138
Figure 4.10   Topography of the high surface resistance specimen
              measured by a focus detection instrument   139–140
                  (a) Isometric plot
                  (b) Grey scale image
Figure 4.11   Topography of the low surface resistance specimen
              measured by a focus detection instrument   141
                  (a) Isometric plot
                  (b) Grey scale image
Figure 4.12   Topography of the superconducting material measured
              by a SEM and an optical microscope   142–143
                  (a) Optical micrograph of the high surface resistance
                      specimen
                  (b) SEM image of the high surface resistance specimen
                  (c) SEM image of the low surface resistance specimen
Figure 4.13   Topography of the fingerprint measured by a focus
              detection instrument   144–145
                  (a) Isometric plot
                  (b) Grey scale image
Figure 4.14   Photograph of a 3-D interferometric instrument
              (WYKO TOPO-3D) in measuring the hip prostheses   146
Figure 4.15   Topography of a successful hip joint measured by an
              interferometric instrument   148
                  (a) Isometric plot
                  (b) Height distribution
Figure 4.16   Topography of a poor hip joint measured by an
              interferometric instrument   149
Figure 4.17   Photography of a STM (Burleigh instrument)   151
Figure 4.18   Topography of a brass surface measured by a STM   152
                  (a) Grey scale image
                  (b) Isometric plot
                  (c) Single trace
Figure 4.19   Isometric plot of the indentation   154–155
Figure 4.20   Intensity plot of the indentation   156

Figure 4.21   Contour plot of the indentation   156
Figure 4.22   Inverted isometric plot of the indentation   157
Figure 4.23   Displaying the pile-up of the indention by combining
               inversion and truncation   158–159
Figure 4.24   Displaying the zoomed area of the indention   159–160
Figure 4.25   Bearing area ratio, material volume ratio and void
               volume ratio of the indention   162
                   (a) Bearing area ratio
                   (b) Material volume ratio
                   (c) Void volume ratio
Figure 1A.1:   Photograph of the ASTA system (courtesy of 3D Digital
               design and Development Ltd, London)   191
Figure 1A.2:   Photograph of the Perthometer S8P (courtesy of
               Feinprüf-Perthen GmbH, Göttingen)   193
Figure 1A.3:   Photograph of the Surfcom 475-3D (courtesy of Advanced
               Metrology Systems, Leicester)   195
Figure 1A.4:   Photograph of the Surfcom 575-3D (courtesy of Advanced
               Metrology Systems, Leicester)   196
Figure 1A.5:   Photograph of the Surfascan-3D (courtesy of Somicronic,
               St. André de Corcy)   198
Figure 1A.6:   Measurement of a cam shaft using the Surfascan-3D
               (courtesy of Somicronic, St. André de Corcy)   199
Figure 1A.7:   Photograph of the Form Talysurf (courtesy of Rank
               Taylor Hobson Ltd., Leicester)   201
Figure 1A.8:   Photograph of the Micromap 512 Optical Profiler
               (courtesy of Burleigh Instruments (UK) Ltd, Harpenden)   203
Figure 1A.9:   Photograph of the Wyko TOPO-3D (courtesy of Wyko
               Corporation, Tucson, Arizona)   206
Figure 1A.10: Photograph of the Wyko RST (courtesy of Wyko
               Corporation, Tucson, Arizona)   208
Figure 1A.11: Photograph of the UBM Optical Surface Measurement
               System (courtesy of UBM Messtechnik GmbH, Ettlingen)   210
Figure 1A.12: Photograph of the Rodenstock RM-600 (courtesy of
               Feinprüf-Perthen GmbH, Göttingen)   213
Figure 1A.13: Photograph of the Maxim 3-D Model 5700 (courtesy of
               Zygo Corporation, Middlefield, CT)   216
Figure 1A.14: Screen display of the Maxim 3-D Model 5700
               topography instrument (courtesy of Zygo Corporation,
               Middlefield, CT)   217
Figure 1A.15: Photograph of the MP2000 (courtesy of Chapman
               Instruments, Rochester, NY)   219
Figure 1A.16: Photograph of the Proscan 1000 (courtesy of Scantron Ltd,
               Somerset)   221

The Contributors

Professor Kenneth J. Stout
Lucas Professor and Head of School of Manufacturing and Mechanical Engineering, at the University of Birmingham (UK) and also heads the Centre for Metrology. He is a mechanical engineer by background. His research areas include surface topography, nanometer technology and precision bearing design. Major areas of activities include the functional assessment of topography and the design of a nanometer measuring environment. He has published over 150 academic papers, edited several books and authored and co-authored four books.

Dr. Weiping Dong
Post-doctoral Research Fellow at the Centre for Metrology in the School of Manufacturing and Mechanical Engineering, at the University of Birmingham (UK), he has experience in signal processing, condition monitoring, machine tool dynamics and computer applications. He is the author of over 35 academic papers and co-authored two books.

Evaristus Mainsah
Research Fellow at the Centre for Metrology in the School of Manufacturing and Mechanical Engineering, at the University of Birmingham (UK). His background is in electronic engineering and his area of research is related to fidelity in topography instrumentation. He has also been involved with the design of laser-based instrumentation for measuring diameter with sub-micron resolution. He has published over 15 academic papers including co-authorship of a book.

Dr. Liam Blunt
Dr Blunt is a Lecturer in the School of Manufacturing and Mechanical Engineering at the University of Birmingham University. His background is in metallurgy, a discipline in which he was awarded a doctorate for his work on grinding. He has also been involved with micro-properties instrumentation design as well as the application of surface topography into Medicine. Additionally, he is involved a number of current research projects within the Centre for Metrology concerning surface topography. He has published 22 academic papers.

Dr. Paul J. Sullivan
Formerly Director of Research in Metrology in the School of Manufacturing and Mechanical Engineering, at the University of Birmingham (UK), he is now with the Precision Engineering Division of the National Institute of Standards and Technology, Gaithersburg. Major areas of activity include topography, micro-hardness and material properties measurement as well as the computer monitoring of precision environments. He has published over 30 academic papers and co-authored two books.

Mr. Naili Luo
Formerly a Research Associate at the Centre for Metrology in the School of Manufacturing and Mechanical Engineering, at the University of Birmingham (UK). Currently a Metrology Engineer with UBM Corporation in Roselle, NJ, USA. His area of research (for which a doctoral thesis is being prepared) is related to filtering and characterisation of surface topography. He has co-authored one book.

PREFACE

Two decades ago, the analysis of 3-dimensional (3-D) surface topography was confined to the academic literature because of the lack of proper instruments for 3-D surface topography measurement. In addition, it was also questionable whether, from an engineering applications point of view, the benefits obtained from 3-D analysis would be significantly greater than those from traditional 2-D analysis. Since then, the growth in demand for the use of quantitative analysis in 3-D surface topography in many disciplines and recent improvements in measurement and analysis techniques have indirectly answered this question.

Scientists working in biology, chemistry and medicine are now no longer satisfied with 2-D photographs taken by microscopes, and the engineers are no longer satisfied with the 2-D profiles obtained from stylus instruments. They need to know the real 3-D surface structure of the observed sample, because of the three dimension nature of surfaces. Increasingly developed measurement and analysis techniques have provided the tools and the approaches required to carry out sophisticated analysis of the surface topography of such surfaces.

However, in spite of the much improved measurement and analysis techniques, there still exists a gap between the available techniques and the acquaintance with the knowledge by engineers and non-specialist researchers. Sometimes, it is difficult to choose a suitable instrument from the range of instruments available for a specific use, and difficult to adopt proper techniques and parameters with which to assess the surfaces.

It is against this background that this monograph has been conceived. It is intended to help bridge the gap between the academic and the engineering applications so as to assist engineers and non-expert researchers in biological, chemical, medical, metallurgical and mechanical disciplines to understand the basic measurement techniques, the commercial instruments in different application fields, the current international situation and perception of 3-D surface topography analysis in academic institutions and industry in Europe, the commonly used 3-D parameters and plots for characterising and visualising 3-D surface topography.

This monograph is composed of four Parts. The first Part reviews the state-of-the-art of instruments and techniques for measurement. Particular attention is given to those which are not normally presented in academic papers, but are commercially available and widely used in science, engineering and industry. Such instruments are the popularly used ones in mechanical and manufacturing industry – stylus instruments; the very promising ones in optical and electronic engineering – optical instruments; and the most effective ones in material and biological science – scanning microscopes. Different from other review papers on surface measurement, the emphasis of this Part is mainly on the process of realising 3-D measurement and its quantitative assessment. The construction and measurement principles of different instruments, their characteristics, including measurement resolution, range, speed and applied conditions, are introduced. Limitations for each technique are discussed from the 3-D point of view. A general overview of these instruments is provided at the end of Part I. An Appendix in this Part gives the technical specifications of some commercial products and the names and addresses of the manufacturers of these products.

The second Part presents the results of a recent survey conducted in relation to an EC project aimed at developing an integrated approach to 3D surface topography assessment. The survey was carried out among surface topography researchers, manufacturers and users in both academia and industry in seven European countries (the UK, France, Germany, Sweden, Belgium, the Netherlands, and Denmark). The academic institutions included universities, polytechnics and research institutes. There was a broad spectrum of industries covered in the survey – instrument manufacturers, car manufacturers, chemical and steel plants and machine tool manufacturers – further evidence that surface topography analysis is multi-disciplinary and now permeates numerous arenas of applied science. The current situations in aspects of surface measurement, reference datum, filtering technique, characterisation methods and parameters are summarised, and the perception for the requirement of functional applications and characterisation are presented. The original questionnaire sheet and the statistical analysis of the questionnaire replies are attached as an appendix to this Part.

To assist the understanding of application of 3-D parameters and visual plots in characterising surface topography, the visual characterisation techniques and a primary parameter set are presented in the third Part of the monograph. Visual plots and manipulation techniques are introduced. The primary parameter set involves a number of useful parameters which were proposed in an EC workshop for 3-D surface characterisation and later distributed widely throughout European industry and academia. They have been tested by characterising many different kinds of engineering surfaces. Full definitions and accessible algorithms of the parameters are provided. The usefulness and effectiveness of these parameters and the visual plots in providing topographic information are established through the use of examples which were carried out on some typical engineering surfaces.

In order to be able to present an overall view of the potential applications of three-dimensional surface metrology, the fourth Part of the monograph introduces some examples of practical uses where recently developed three-dimensional surface measurement and characterisation techniques have been used. All applications were carried out at the Centre for Metrology at the University of Birmingham employing a comprehensive range of instruments including stylus, optical focus detection, optical interferometry and the scanning tunnelling microscope. These applications cover not only conventional engineering areas, but also non-traditional areas such as bio-engineering and physics. The performances of the instruments are discussed in terms of the above examples. The system set-ups are illustrated through the use of photographs of the instruments in use. A number of 3-D characterisation techniques are used to analyse the measurement results.

Part V is a comprehensive bibliography comprising more than 30 references in surface topography and related disciplines that have been consulted in the course of this work. Included in the bibliography are technical Papers, reference texts, patents and official reports. These range from the earliest Papers reporting major pioneering work, through patents exploiting new techniques, to Papers reporting applications in new areas. The references go as far back as the period when the earliest topography instruments were conceived and cover material up until the end of 1993 and should be of great use to people who only require general information as well as those with more than just a passing interest in the subject.

It is hoped that this monograph will go some way towards assisting the understanding of 3-D surface topography analysis and that many more branches of science, engineering and medicine will, as a result, be able to take advantage of its scope and power.

ACKNOWLEDGEMENTS

The authors would like to thank the past and present members of the Digital Surface Analysis Research Group (DSARG) of the Centre for Metrology, School of Manufacturing and Mechanical Engineering at the University of Birmingham for their support during the production of this Monograph.

The authors would also like to thank the following instrument manufacturers (or their agents) who cross-checked their instrument specifications and also supplied photographs which have been used in Appendix 1A of Part I –

3D Digital Design and Development Ltd, London, (*3-D ASTA*);

Feinprüf-Perthen GmbH, Göttingen, (*Perthometer S8P*).

Advanced Metrology Systems, Leicester, (*Surfcom 475-3D/575-3D*).

Somicronic, St. André de Corcy (Surfascan-3D).

Rank Taylor Hobson Ltd., Leicester (*Form Talysurf*).

Burleigh Instruments Ltd, Harpenden, *(Micromap 512 Optical Profiler)*.

Wyko Corporation, Tucson, AZ (*Wyko RST, Wyko TOPO-3D*).

UBM Messtechnik GmbH, Ettlingen (*UBM Optical Surface Measurement System*).

Feinprüf-Perthen GmbH, Göttingen (*Rodenstock RM-60*).

Zygo Corporation, Middlefield, CT (*Maxim 3-D Model 5700*).

Chapman Instruments, Rochester, NY, (*MP200*).

Scantron Ltd, Somerset, (*Proscan 1000*).

Part II of this work benefited from the support of the Commission of the European Communities who part financed it under its Programme of Applied Metrology and Chemical Analysis – BCR No. 3374/1/0/170/90/2. This help is acknowledged.

The results reported in Part II are as a consequence of the many people in industry as well as academia in several EC countries who took the time to respond to the questionnaire. The authors are indebted to them for this.

Finally the authors are particularly indebted to their editors at Jessica Kingsley Publishers who have worked so hard over the last few months to make the manuscript readable.

K.J. Stout
Birmingham
January 1994

Part I

INSTRUMENTS AND MEASUREMENT TECHNIQUES OF 3-DIMENSIONAL SURFACE TOPOGRAPHY

W P Dong, E Mainsah,
P J Sullivan and K J Stout

INSTRUMENTS AND MEASUREMENT TECHNIQUES OF 3-DIMENSIONAL SURFACE TOPOGRAPHY

In recent years, quantitative measurement of 3-dimensional (3-D) surface topography has been increasingly applied in many science and engineering fields. In this Part the state-of-the-art in instruments and techniques for the measurement is reviewed. Particular attention is given to those instruments which are not normally presented in academic papers, but are commercially available and widely used in science, engineering and industry. Also included are the very promising instruments available in optical and electronic engineering – optical interferometry, and the most effective ones used in material and biological science investigations – scanning electron and scanning tunnelling microscopes. Different from other reviews on surface measurement, the emphasis of the work described is mainly concentrated on the mechanism of realising 3-D measurement and its quantitative assessment. Constructions and measurement principles of different instruments, their characteristics, including measurement resolution, range, speed and applied conditions, are introduced. Limitations for each technique are discussed from the 3-D point of view. A general overview of these instruments is included at the end of Part I.

1.1 INTRODUCTION

It is well known[1,2] that surface topography greatly influences not only mechanical and physical properties of contacting parts, but also optical and coating properties of some non-contacting components. The characteristics of surface topography in amplitude, spatial and pattern of surface features dominate the functional applications in the fields of wear, friction, lubrication, fatigue, sealing, jointing, reflecting, painting, bearing surfaces, optical properties etc. In addition, the tomography of organs, tissues, cells, geological fossil, metallic and non-metallic materials are also of interest to biologists, chemists, geologists and metallurgists. Measurement and analysis of surface topography has therefore attracted attention and as a consequence is now winning increased interest from both industry and academia.

Recently, the measurement of surface topography has evolved in two promising directions: precision measurement and three-dimensional measurement. The former has led to the possibility of atomic size measurement (on the ångström scale of magnitude) and the latter has resulted in a dramatic change of measurement of surface topography from a regime where 2-dimensional (2-D) profiling and sectioning techniques had been used for nearly 60 years and is still going to be widely used in the foreseeable future, to a situation in which quantitative 3-D surface topography is obtained.

It is accepted that surface topography is three dimensional in nature; and any measurement and analysis of 2-D profiles or sections, even if properly controlled, will therefore give an incomplete description of the real surface topography. Fundamentally, only 3-D quantitative measurement (the term 'quantitative measurement' refers to 'profiling' or 'non-parametric measurement',[3–5] by which the surface topography of a specimen is obtained in terms of the variation in its height as a function of position x & y) can give a complete description of surface topography. The information to be provided by 3-D measurement is far more comprehensive than 2-D profiles or sections previously available.

1.1.1 3-D, Areal, and Parametric Measurement

The term *3-D measurement* in this context refers to *2-D profile measurement* e.g. by stylus instrument, *areal measurement* e.g. by a light microscopy and an ordinary SEM, and *parametric measurement* e.g. by light scattering instruments. The technique provides quantitative 3-dimensional information of measured surface topography, while other measurements can only provide quantitative 2-dimensional information (either one vertical axis and one horizontal axis or both horizontal axes) or parametric information. Areal measurements and parametric measurements have the sense of 3-D measurement since they obtain information from 3-D topography; however, they cannot provide quantitative information of surface topography with respect to position. The two techniques actually provide pseudo three-dimensional measurement.

1.1.2 History of Surface Measurement

The practice of areal measurement could be traced back to about three hundred years ago when the first microscope was invented.[6-8] Surface topography was viewed through the microscope, but no quantitative surface height information could be obtained. *Optical specular reflectance* was also an early areal measurement technique[9] used, but it provided neither quantitative surface topography nor an image of the surface. However, a parameter which represented root mean squares (RMS) roughness was presented. Significant progress in areal measurement was made about the same time when the first *stylus profilometer* was invented, as a consequence of the development of the first *transmission electron microscope* (TEM)[10–14] and the *scanning electron microscope* (SEM)[15–18] in the 1930s. The early TEM and

SEM had horizontal resolutions of hundreds of ångström. To this day they remain non-quantitative measurement techniques.

Although some areal measurement instruments, such as the *interference microscope*[19–22] and the *scanning transmission electron microscope* (STEM),[13,14,18] were developed subsequently, quantitative measurement of 3-D surface topography remained less than fully satisfactory until the middle of 1960s. This was partly due to a lack of appropriate measurement approaches, and partly due to the limitations in development of digital computer techniques. Since then, 3-D measurement techniques have developed rapidly as digital computer technology and other related measurement techniques developed. At the end of the 1960s, initial stylus 3-D measurement systems were proposed by Williamson, Peklenik *et al.*[23–26] The work of Sayles and Thomas[27] rendered such a system of more practical use in 1976; A further instrument – the *topografiner*, which is based on the principle of filed emission, was originated in 1966[28] and completed in 1972 by Young.[29] Quantitative SEM, in which a stereo pair is used to calculate surface height, was developed around the early 1970s.[30–34]

One of the currently commercialised optical instruments for 3-D surface topography measurement is called the *focus detection instrument*.[35–38] Earlier instruments of this kind were proposed by Minsky[39] in 1957 for area measurement and by Dupuy[40] in 1968 for the measurement of 2-D profiles. Several techniques[41–49] of *focus detection* were developed around the early 1980s, and 3-D measurement was introduced into this technique.[50–52] A promising optical instrument for the quantitative measurement of surface topography is the *interferometer*. Many quantitative profiling techniques[53–75] based on optical interference theory were developed around the 1980s. Some of these techniques[66–69,74] have been successfully applied to the measurement of 3-D surface topography since the middle of the 1980s, and have subsequently become the forerunners of current commercial systems.[76–79]

A breakthrough in 3-D precision measurement techniques was made by Binnig and Rohrer with the invention of the *scanning tunnelling microscope* (STM)[80–82] in 1981, and the subsequent invention of the *atomic force microscope* (AFM)[83] in 1986. The significance of these two instruments is that they have nanometer or sub-nanometer resolutions in both lateral and vertical directions so that they can detect topography features on an atomic or molecular scale. The STM and the AFM are quantitative measurement instruments, which are used together to measure the micro topography of conducting and non conducting materials.

Since advances in quantitative measurement instruments in 3-D surface topography have occurred, it is no longer difficult to obtain the topography of a practical surface over different scales ranging from atomic to machined components. However, in order to understand some basic principles and characteristics, it is of interest to review applications and developments in measurement instruments for 3-D surface topography. It is helpful for mechanical engineers, tribologists, chemists and biologists to be in a position to properly choose and apply the necessary instruments or to design their own measurement systems when necessary. Therefore, this Part aims to present a review of current widely used measurement instruments for surface topography analysis. Unlike other reviews[3–5,84–96] of the measurement of surface topography, this Part is specially concerned with the quantitative measurement techniques of 3-D surface topography. Some quantitative profiling techniques[56,97] which are not used for 3-D measurement, and some non-quantitative area techniques[98–108] which measure surface topography to obtain parametric or visualisation information are not included in this Part. As such, this Part emphasises the mechanisms of realising 3-D and quantitative measurement, constructions and measurement principles used by different instruments. Characteristics, including measurement resolution, range, speed and applied conditions, are reviewed. The problems encountered in each technique are discussed from the 3-D point of view. This review covers measurements ranging from objects as small as biological cells to routinely used optical and engineering surfaces. A general survey of these instruments is given at the end of this review.

1.2 DIFFERENCES IN THE MEASUREMENT AND ANALYSIS METHODS FOR 2-D AND 3-D SURFACE TOPOGRAPHY

As has been previously stated, 2-D profile measurement and its analysis are still playing an important role in the assessment of surface topography. The positive features of the technique are the short time needed to measure a profile, the ability to provide some (not comprehensive) information about surface topography and the lower cost of the instrument. Thus, this technique is currently preferred in industry to 3-D measurement techniques and this will probably continue to be the case, unless the cost and the time of 3-D measurements can be reduced to a level comparable with 2-D measurements. In spite of this, evidence from a recent questionnaire[109] seems to show that the 3-D measurement of surface topography is not simply of academic interest because many industrial companies have started to use 3-D techniques to assess the topographies of their products. The main reason for this activity is probably due to the fact that 3-D measurement is able to provide complete information about surface topography.

Physically, 3-D topography measurement is not complex when compared with 2-D profile measurement; only one more dimension is added to a previous 2-D measurement system. However, the intrinsic benefits of 3-D measurement and analysis of surface topography go far beyond the simple physical realisation that can be achieved from 2-D analysis. Briefly, 3-D analysis has the following main characteristics compared with 2-D analysis.

(1) It is known that surface topography is three dimensional in nature. An obvious characteristic of the measurement of 3-D surface topography is that it can represent the natural characteristics of the surface topography, whilst the measurement of 2-D profile does not achieve this. The limitation of 2-D assessment is shown in Fig. 1.1(a). As is seen from Fig. 1.1(a), which is a profile from a ground

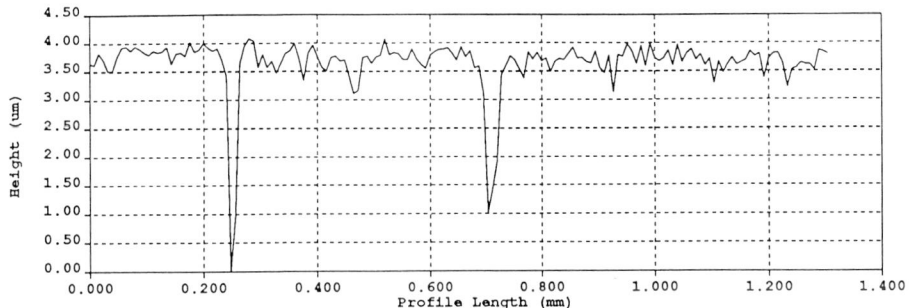

(a) Profile of the ground surface

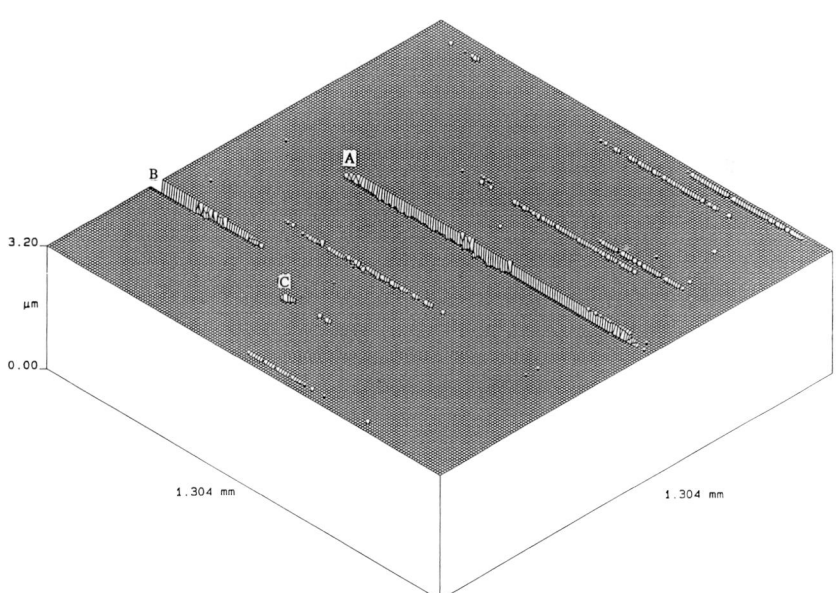

(b) Isometric plot of a truncated ground surface

Figure 1.1 2-D and 3-D plots of a ground surface

surface,[110,111] two deep valleys exist on the profile; it is impossible to identify whether the valleys are from pits or troughs by this representation of the surface. However, by 3-D characterisation of the surface topography, an isometric plot (Fig. 1.1(b)) of the truncated surface topography presents a clear identification of pits and troughs existing on the surface. Moreover, not only are the qualitative identification of surface features, such as pits, troughs, lay, anisotropy, and inhomogeneity easily obtained from the 3-D characterisation of the surface topography, but also the sizes, shapes and volumes of the features can be calculated quantitatively.

(2) Parameters obtained from 3-D surface topography are more realistic than those obtained from 2-D profiles. Some extreme parameters,[112–114] such as R_p, R_v, R_y, R_t, R_z are provided by 2-D profile characterisation. Since the profile coincides with an intersection of a vertical plane with the measured surface, it may not cross the real surface summits or valleys. The extreme parameters obtained from the profile, therefore, are only approximate indications of the real values. However, by the measurement of 3-D surface topography, true summits and valleys can be found. Thus the extreme parameters are more realistically represented in this case. A similar situation also occurs when calculating the bearing ratio curve[112–114] – a profile bearing ratio curve may be significantly different from the area bearing ratio curve since a profile does not have a true meaning of 'bearing area', especially for a profile obtained from a random surface.

(3) The measurement of 3-D surface topography can provide some new meaningful parameters, such as oil volume, debris volume and contact area[115] which are not available from the conventional analysis of 2-D profiles. These extra parameters are very useful for engineers and tribologists, enabling them to attempt to analyse the functional properties of engineering surfaces.

(4) From a statistical point of view, the more independent sampling data available, the better the evaluation of the ensemble property of the random process. Thus, the statistical analysis of 3-D surface topography is more reliable and more representative since the large volume of data obtained using 3-D topography increases the independence of the data. This is especially true for stationary random surfaces, where the increment is more significant.[116–119] In other words, 3-D analysis can reduce the variance of parameters, especially for those parameters calculated from theoretically stationary

surfaces such as produced by *electro-discharge machining* (EDM) and shot blasting surfaces.

(5) One of the most important characteristics of 3-D topography analysis is that a visualisation technique is provided through the use of a computer. With the help of computer and image processing techniques, any 3-D non-optical measurement system can produce images which are equivalent to those taken by a microscope. Fig. 1.2 shows a grey scale map logged from a coin with the 3-D stylus system developed at the University of Birmingham. Each 3-D data point is represented as a level of grey (between 0 and 255) whose intensity is proportional to the magnitude of the height. This type of picture looks similar to a photograph and is able to represent a real surface with a picture quality similar to that viewed through a microscope. Other kinds of visualisation pictures are also available.[120] Isometric projection, contour map, inversion and truncation pictures provide detailed surface topography, allowing different aspects to be visualised. The information obtained from them is significantly greater than can be achieved by 2-D profile measurement.

Figure 1.2 A grey scale map of a coin logged by a 3-D stylus system

(6) Many 2-D measurement systems still in use are analogue systems, especially those developed for use in workshops, whereas almost all quantitative 3-D measurement systems are digital systems. The advantages of digital systems are that they have powerful functions to manage data as required by the users. Such systems are flexible in their ability to process data, convenient to store the data for frequent or permanent use and convenient to operate through user-friendly menus, windows and soft keys.

Although the measurement and analysis of 3-D surface topography has such promising characteristics, it has some disadvantages. The cost of instruments and the time of measurement are serious drawbacks which prevent the 3-D technique gaining widespread popularity. Nevertheless, 3-D techniques are still expected to gain increasing use not only in the areas where mainly 3-D measurement can get meaningful results, such as biology and chemistry, but also in the areas where traditional 2-D techniques have been conventionally applied, as in optical and mechanical engineering.

1.3 STYLUS INSTRUMENTS

It was over thirty years after the invention of the first stylus instrument that 3-D measurement using stylus instrument was proposed by Williamson, Peklenik et al. [23–26] It is believed that the first computer-controlled stylus 3-D system was completed by Sayles and Thomas[27] in 1976; since then, a large number of 3-D stylus systems have been built[121–135] and some systems are now commercially available.[136–139] Despite some disadvantages,[140–147] stylus instruments are most popularly used to measure 2-D profiles of engineering surfaces and to define national and international standards.[148,149] Similarly, stylus 3-D systems are versatile, reliable and easy to construct from corresponding 2-D systems. They have some different characteristics compared with the 2-D systems. The measurement mechanism, the construction and the features of stylus 3-D systems are therefore discussed here.

1.3.1 Mechanism of 3-D Profile Measurement

The mechanism of 2-D profile measurement by stylus instrument has already been described by many authors.[1,2,150,151] A stylus which has a very small tip is traversed (say direction X) across the surface to be measured, and a pick-up, which is either a linear variable differential transformer (LVDT) or an optical transducer, physically connected to the tip, converts the vertical movement (say direction Z) of the stylus, into an electrical signal. The electrical signal is amplified and processed by subsequent electronic circuits. In order to realise 3-D measurement, one more dimension is needed in the measurement. Generally, there are two methods to realise the third dimension of movement.

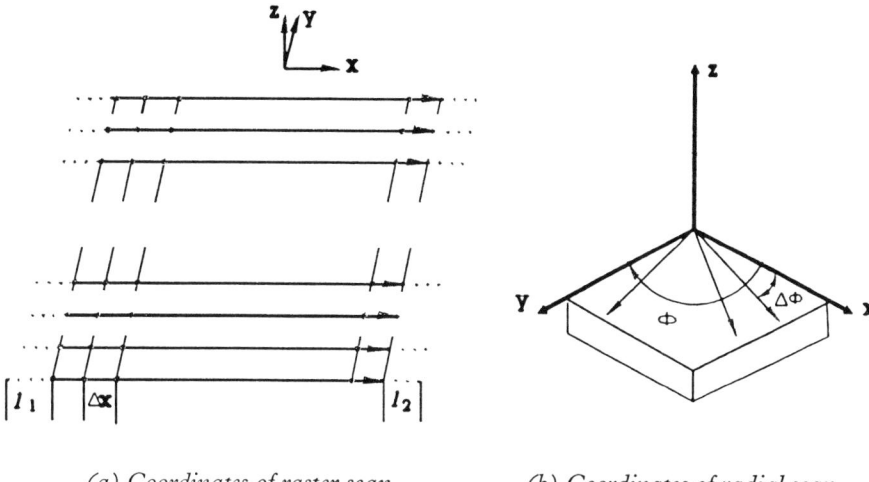

(a) Coordinates of raster scan   (b) Coordinates of radial scan

Figure 1.3 Coordinates of scan modes

(1) *Raster scan*. This method takes a number of closely spaced parallel profile traces which are referenced to a common origin, to give the third dimension. As shown in Fig. 1.3(a), the first two dimensions, X and Z, are realised when a 2-D profile is traced, whereas the third dimension Y, which is perpendicular to the X-Z plane, is generated as each trace scan is obtained. When the system starts to scan each trace, the origins must be kept in the same Y-Z plane. Since 3-D systems normally achieve their logged data by digital sampling, the sampling intervals ΔX and ΔY are limited. Generally, the same value for ΔX and ΔY is selected for collecting and displaying the data. However, if the statistical properties and the inhomogeneity of the surface topography are of great interest, it is also possible to select ΔX and ΔY with different values. For example, for increasing the independence of the parallel profiles and for observing the characteristics of inhomogeneity, a larger ΔY may be useful. Fig. 1.4 shows an analysis result of such a selection. A milled surface is presented with $\Delta X=8$ μm and $\Delta Y=80$ μm; the inhomogeneity of the milled surface is clearly visible. The raster scan method suits the analysis of all kinds of surface topographies and can easily be con-

structed by commercially available components. It is not surprising, therefore, that almost all current 3-D stylus instruments[27,121–139] are based on this method.

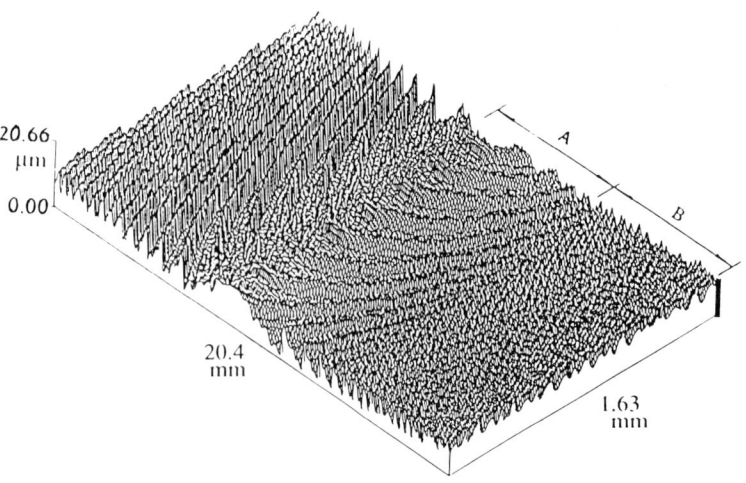

Figure 1.4 An isometric plot of an inhomogeneous surface logged using different sampling interval in X and Y directions

(2) *Radial scan.* Unlike the raster scan method, the profiles are not collected by located parallel traces, but are taken with respect to radial angles. As is seen from Fig. 1.3(b), the first two dimensions, R and Z, are realised in the same way as the raster scan, but all traces start from the same position. The third dimension is a radius denoted by Φ, and successive profiles are traced with differential angle $\Delta\Phi$. This method was proposed by Peklenik and Kubo[24,25] and was sometimes used to analyse characteristics of isotropic and anisotropic surfaces with a few angular distributed profiles.[229–231] Since the visual representation provided by this method is not as trivial as that provided by a raster scan for anisotropic surfaces, and due to the difficulties of physical realisation among other reasons, the method is rarely used in current 3-D stylus data acquisition systems.

Regardless of the scan method selected there are two possible types of data collection modes, that is, there are two kinds of profile digitising processes. The first one is called *dynamic* or *on-the-fly measurement*, by which data sampling is synchronous with the stylus scan. This mode can reduce data logging time. According to the investigation recently carried out at the University of Birmingham stylus speed can be as high as 4 mm/s, in most cases without significant effect on measurement results. However, the stability of the speed is an important factor affecting position accuracy of data collection. The second mode is called *static (point to point) measurement*. In this case, the stylus stops for every traverse length ΔX, and then the system obtains an individual piece of data. This scanning process is very slow when compared with dynamic acquisition because of the interruptions that occur when collecting data. However, the sampled data may be more reliable, and not be influenced by the speed and dynamic characteristics of the stylus.

For all lead screw driven translation stages, a common problem is that backlash of the mechanical mechanism is inevitable. Thus an excess traverse length, l_1, is usually required at the beginning of each traversing as shown in Fig. 1.3(a). This distance is used for removing backlash and has been set at $100\,\mu m$ by some authors.[121] Under the dynamic measurement mode, the distance l_1 should be longer than in static measurement to ensure that constant velocity is reached, and another excess traverse length, l_2, is required at the end of each traverse to decelerate. This is to ensure that the data sampling process should occur within the constant traverse speed range as shown in Fig. 1.5.

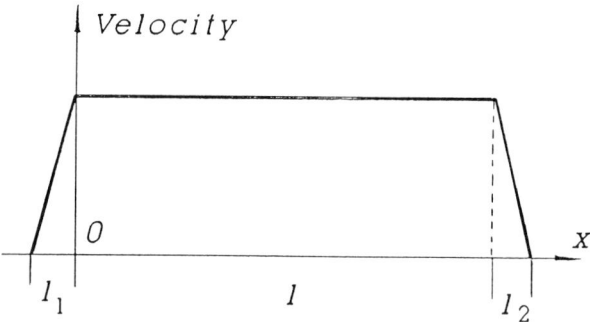

Figure 1.5 Plot of traverse velocity versus position

1.3.2 Construction of the System

The constructions of a conventional 2-D analogue system and two digital 3-D systems are shown in Fig. 1.6. The major difference between the 3-D system and the 2-D system is that one more stage is used in the former. The computer used in the 3-D digital system is the management centre which controls all aspects of the measurement processes. The stage operation command is transmitted to the motor driving units through a serial or parallel interface. The interface is dual in direction, thus handshaking and communication between the motor driving units and the computer is required.

It is evident, from Fig. 1.6(b) and (c), that the difference between the structures of the two 3-D systems is the X translation mode. The X translation in Fig. 1.6(b) is realised by the use of gear box, and a stage is used to realise X translation in Fig. 6(c). In both cases, the stage or the gear box is driven either by a stepper motor, a DC motor or a linear motor, which is controlled by a driving unit. This difference in the mechanism introduces different measurement datums. In the system shown in Fig. 1.6(b), the measurement datum is achieved by referencing the output against an optical flat under the pick-up and the Y stage, whilst the measurement datum of Fig. 1.6(c) is a combination of the movement of the X and Y stages. The movement errors of the X and Y stages would induce the error of the datum.

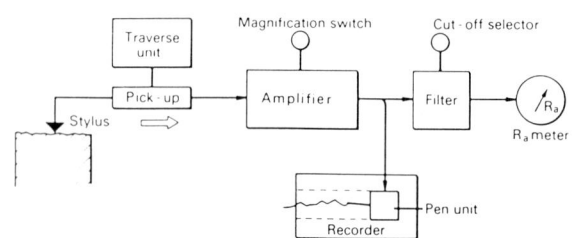

(a) A conventional 2-D system (from Dagnall) [2]

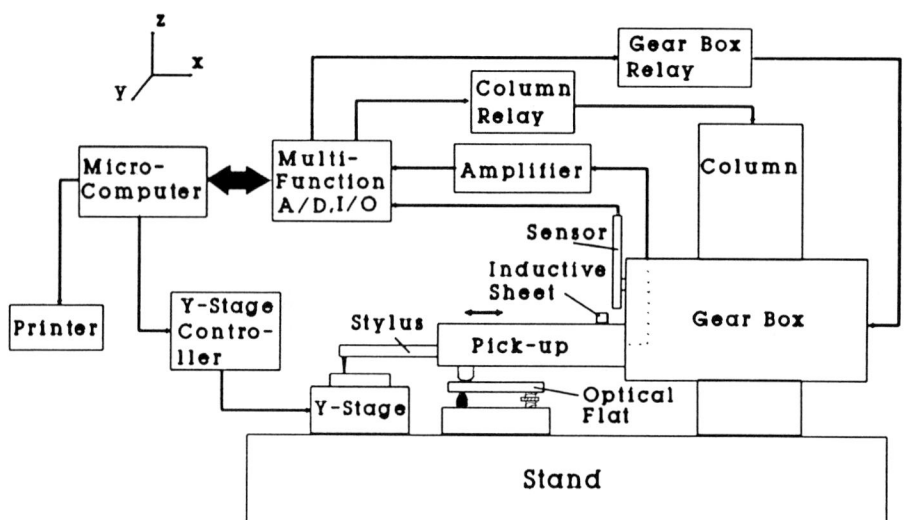

(b) A 3-D system with one translation stage and a gear box

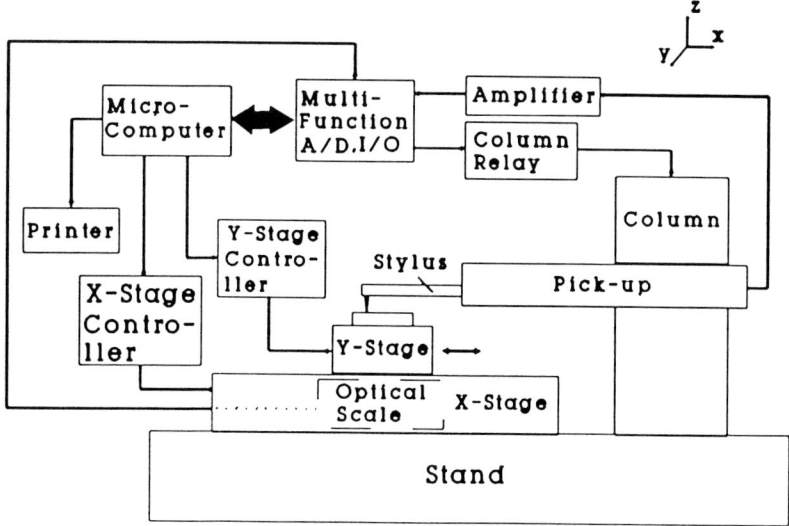

(c) A 3-D system with two translation stages

Figure 1.6 Schematic diagrams of 2-D and 3-D stylus systems

As the stylus scans the surface, the pick-up (the mechanism of the pick-up has been described elsewhere[1,2]) converts the mechanical movement of the stylus to an analogue signal which is transmitted to an amplifier and then through an analogue filter. In 2-D profile measurement, the filter is usually a high-pass filter whose cut-off length is defined in national and international standards.[148,149] However, in 3-D surface topography measurement, in order to maintain the same datum for all measured traces, the filter should be an all-pass-band filter or alternatively no filter is used at all so that the true profile can be obtained. The necessary filter for topography characterisation can be implement digitally after the analogue signal is digitised by the analogue to digital (A/D) converter.

An important feature of 3-D measurement is that the start points of all scan traces should be kept in the same Y-Z plane as shown in Fig. 1.3(a). In other words, it is vital to relocate the start point of each trace so that multiple parallel traces can be taken from the required area. Furthermore, data sampling should be active at every movement ΔX of the stylus (Fig.1. 6(b)) or the X stage (Fig. 1.6(c)). Several data acquisition approaches can thus be adopted.

(1) *Open loop position trigger.* This approach is mainly used for static measurement. Data acquisition position is determined by the number of pulses sent to the stepper motor. For example, if each pulse corresponds to a 1.8 degree rotation of the stepper motor, and the lead screw pitch is 0.5 mm, then 200 pulses are required to produce 0.5 mm traverse length, and the minimum spatial resolution is 2.5 μm. Since there is no feedback position control with this approach, the position accuracy is totally dependent on the mechanical driving characteristics of the stepper motor and the lead screw, and the dynamic characteristics of the stepper motor.

(2) *Open loop timing trigger.* An example is shown in Fig. 1.6(b). An approach sensor is used to produce the start trigger pulse and an internal timer within the A/D board produces a timing signal to trigger the data acquisition. This approach can have satisfactory start points of data acquisition for the parallel profiles and is suitable for the dynamic measurement. However, the positioned accuracy within each scan is influenced by the inconsistency of measurement speed and the accuracy of the internal timer.

(3) *Closed loop position trigger.* As in the example shown in Fig. 1.6(c), an optical scale is used for sensing the stage position and producing the trigger signal. In this case, positional accuracy for data acquisition is neither influenced by the instability of the measurement speed nor by the inaccuracy of the timer; exact data acquisition positions can be obtained, theoretically. However, in reality, the positional accuracy is influenced by the number of lines of the optical scale and their accuracy. This approach is highly suited to dynamic measurement.

One component of the 3-D system which is not indicated within Fig. 1.6 is a levelling device. It is known that if a surface is not levelled the measurement and characterisation may be greatly distorted. There is a reduction of the lateral and vertical measuring range and the reliability of the data becomes questionable. Therefore, it is important to level specimens with a levelling device in 3-D assessment systems. Currently there are two main kinds of levelling devices; one is motor driven[134] and the other is manually adjusted. Due to the simplicity of the structure, the latter is more widely used in practice.

In some 3-D systems, a mechanism which lifts the stylus for the return traverse is adopted.[152] This is introduced to increase the return traverse speed hence to reduce the surface mapping time. It also avoids producing a 'skipping action' or ploughing and damaging of the stylus tip.

1.3.3 Stylus 3-D Systems

Since 3-D stylus systems are not, in principle, radically different from 2-D stylus systems, almost all problems encountered by the 2-D system do appear in the 3-D system. These problems, which involve the effects of stylus size, shape, skid, load, deflection and dynamic characteristics, have been discussed in the literature.[140–147] However, some still need to be considered in detail when approaching 3-D measurement.

(1) *Effect of stylus size.* When measuring very fine surfaces, stylus size is a critical factor which influences measurement results. Although these influences have been discussed by some researchers, the discussion was mainly concentrated on two dimensional problems, i.e. the stylus was regarded within a vertical plane. In reality, the stylus is three dimensional and therefore does not only suffer from interference with the measured surface along the trace direction, but also sees interference along the orthogonal direction. Well defined lay structured surfaces (*anisotropic surfaces*) and lay free structured surfaces (*isotropic surfaces*) may have the same interference with the stylus along the trace direction, but the interferences along the orthogonal trace direction would be totally different, especially, when the sampling interval ΔY is very small compared with the tip radius. As a consequence, the effects of the stylus must be considered further.

(2) *Effect of dynamic characteristics of the stylus.* Although this effect has been discussed in some papers, the results obtained are not conclusive. Little by way of experimentation has been carried out to test the dynamic characteristics of the stylus, i.e. to test the damping ratio and natural frequency of the stylus. No optimum speeds for dynamic measurement have been suggested for various kinds of surfaces. The speeds recommended for 2-D measurement by manufacturers of instruments are usually too conservative, so prolonging 3-D data acquisition time. Further research on this problem needs therefore to be carried out.

(3) *The effect of skid.* As is well known, skid influences the measurement of 2-D profiles. Obviously, the effect would be more severe in 3-D measurement. Since 3-D measurement is more concerned with the common datum than 2-D measurement, the influence of skid has to be taken out of the measurement. The easiest way to do this is to remove skid in 3-D measurement and to let the assessment datum be solely the movement of X-Y or the combined movement of the traverse unit and the Y stage as mentioned in the last section.

(4) *Co-ordinate error*. Besides the error induced by the stylus, co-ordinate error is another serious error distorting the measurement result. This has been discussed by Sullivan[153] and Sherrington,[154] and mainly arises from the error of construction of mechanical and electronic driving devices, such as stepping or DC motors, gears, bearings and lead screw, all of which affect positioning, relocation of parallel traces and stability of the measurement speed significantly. This error appears in all three dimensions, so the total error in each sampling point is a vector superposition of the three individual errors. It is significant in the measurement of fine surfaces, and exists in both static and dynamic measurement modes. Since it is a systematic error, it can to some extent[153] be compensated for.

As for the range and resolution of stylus 3-D systems, the vertical range is similar to 2-D systems, and is determined by the pick-up employed. Most pick-ups are LVDTs, optical-electrical or piezo-electric components. The vertical range is about 0.5 mm – 1 mm, with some such as the Form-Taylor-Surf, Perthometer S8P and Surfascan-3D having a range of 4–6 mm.[137,138] The vertical resolution of 3-D systems is similar to 2-D systems. A minor difference of a 3-D digital system when compared with a 2-D analogue system is that the vertical resolution of the 3-D system is also related to the bits of the A/D converter. The longer the bits, the smaller the quantization error and the higher the resolution that can be obtained. Combining the pick-up, the magnification of the instrument and the A/D converter, the highest vertical resolution can be in the ångström scale.[87,109] Currently most stylus 3-D systems use 12 bits, although 8, 13, 14, 15 and 20 bits are also available.[109] When considering the ratio of characteristics against price, the 12 bit A/D converter seems to give the best value for money.

The spatial range of stylus 3-D systems represents an insignificant restrictive feature in its use. A general stylus 3-D system has a wide measurement range from 50x50 to 150x150 mm, which is large enough for the measurement of surface roughness, because the areal measurement of the micro topography of surfaces is normally within a square with side lengths of only a few millimeters. Sometimes the spatial range is also referred to as the 'number of sampling data points'. From the analysis of the questionnaire,[109] an area of about 128x128 data points is frequently adopted to obtain an estimate of the surface topography and is in routine use, while an area with 256x256 data points or more is usually adopted in academic research. The maximum number of sampling points is restricted by the capacity of the computer memory (normally it is 640K). Although extended memory and hard disk can be used to manipulate an areal data of more than 512x512, it is often very tedious.

The spatial resolution of stylus 3-D systems is similar to that in 2-D systems; the highest resolution is limited by stylus shape and size, the resolution of stepper motors and the pitch of the lead screws. The resolution is also dependent on the topography of the measured surface. It is easier to measure comparatively large spatial topographies than to measure small spatial topographies. Thus it is difficult to define the exact spatial resolution for stylus instruments. The declared highest spatial resolution is 0.1 μm.[87] In some cases a pseudo highest spatial resolution can be obtained with an open loop timing trigger and close loop position trigger. In the first case this is realised by increasing the timing frequency and reducing the measurement speed. In the second case an electronic divider circuit is used to create the trigger pulses between two neighbouring lines of the optical scale. Due to the limitation of stylus shape and size, very higher spatial resolutions do not make a significant contribution to the measurement result, because the sampled data is like to be highly correlated.

1.4 OPTICAL INSTRUMENTS

Due to the need for fast, non-contact and versatile measurement of surface topography and also the advent of modern optical-electronic and digital computer techniques, optical instruments have developed very quickly. If it is true to say, about ten years ago, that optical instruments were mainly concentrated on areal and parametric measurements, then 3-D measurement and profiling techniques presently dominate the research of optical instruments for surface topography measurement. Based on various optical principles, many optical profiling instruments[39–52,60–69,72–75,155–158] for surface topography measurement have been proposed. Among them focus detection instruments[39–52] and interferometers[60–69,72–75] are most prominent. They are commercially available and widely used in mechanical, manufacturing, electronic, optical, geological, chemical, and biological areas. The principles and constructions of the two kinds of instruments are now outlined.

1.4.1 Focus Detection Instruments

The term *focus detection* is a very general concept; it implies that the surface profile is measured by maintaining the focus of the optical system or making use of the principle of optical focus. Early reports on focus detection techniques used to measure the tomography section and 2-D profiles were made by Minsky[39] and Dupuy.[40] Since the early of 1980s, 3-D measurement using the focus detection technique have been introduced,[50–52] and now focus detection instruments have been increasingly accepted for engineering applications and academic research,[35–52] especially in biological research, where one focus instrument – the *confocal laser scanning microscope* (CLSM)[159–176] – has become routinely used for observing the tomography of cells and tissues.

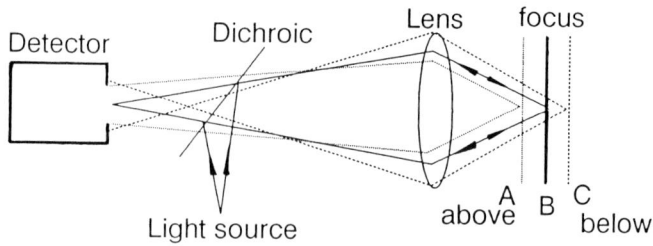

(a) A general focus detection system

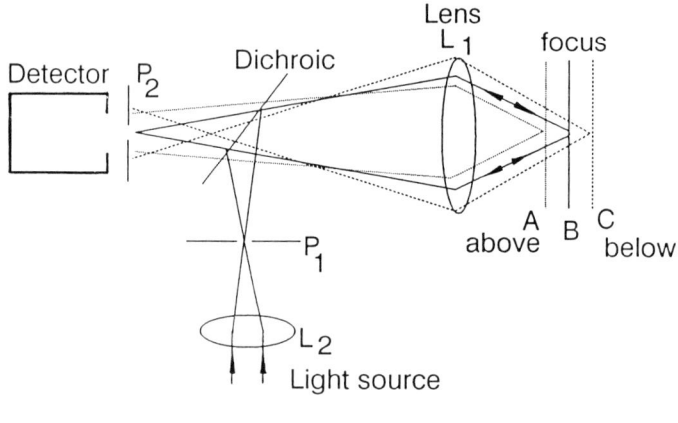

(b) A confocal system

Figure 1.7 Optical arrangements of focus detection systems

The principle of focus detection for surface measurement is simple. As is seen from Fig. 1.7(a), an illuminating light is reflected by the dichroic mirror, and is focused by the objective to a diffraction limited spot at the focus plane B. Since the diameter of the spot is very small ($\approx 1\ \mu$), 3-D measurement is implemented by a raster scan of the light spot over the measured surface. If the focused spot can be kept on the top of the surface topography by adjusting the objective or the specimen vertically in the scanning process, or in other words, if the top point of the surface topography being measured is a constant spot in the focus plane in the scanning process, then the vertical dimensional information of the surface topography is determined by the movement of the objective or the specimen. Since the key problem of the focus detection technique is to detect focus as sensitively and conveniently as possible, or to make use of the principle of optical focus, many types of focus detection methods have been developed.[39–52] These methods differ mainly in (i) focus detection mechanism, (ii) vertical scan method and (iii) horizontal scan method. Some main focus detection methods are summarised below.

1.4.1.1 Intensity Detection Method[5,42]

The intensity focusing method is based on the principle that if a point on a surface is coincident with the focused light spot (position B in Fig. 1.7(a)), the maximum power of the reflected light can be obtained. An intensity focusing system is shown in Fig. 1.8; a collimated laser beam traversing a beam splitter BS is focused onto the surface of the investigated sample by an objective L_1. The light reflected from the surface returns through L_1 and is directed by the beam splitter BS to an objective L_2. A photodetector D is located at the focus plane of L_2 to collect the reflected light and to convert it to an analogue signal. When the system scans horizontally, the surface height under the light spot changes, so by adjusting the relative position of the objective L_1 and the sample, (in the Fig. 1.8, the lens L_1 is moveable), the focused light spot can be kept on the surface whilst the intensity of the light transmitted to the photodetector changes as the relative position changes. The output signal from the photo detector is inputted to the control unit to form

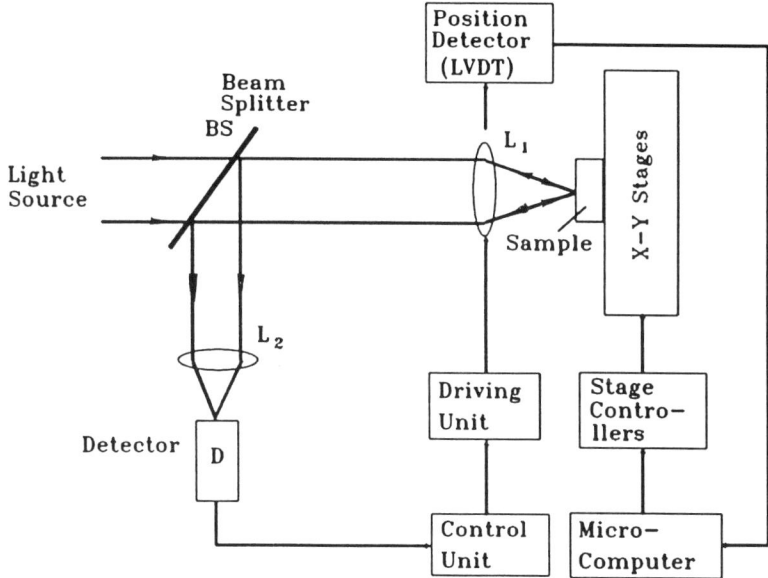

Figure 1.8 Schematic diagram of an intensity focus detection system

a feedback control process for maintaining the relative position between the objective L_1 and the sample by driving a linear motor or a piezo-electric translator. Changes in the position of the objective L_1, which represent the measured topography, are sensed independently of the focus control system by an LVDT. This guarantees a linear measurement signal throughout the entire measurement range.

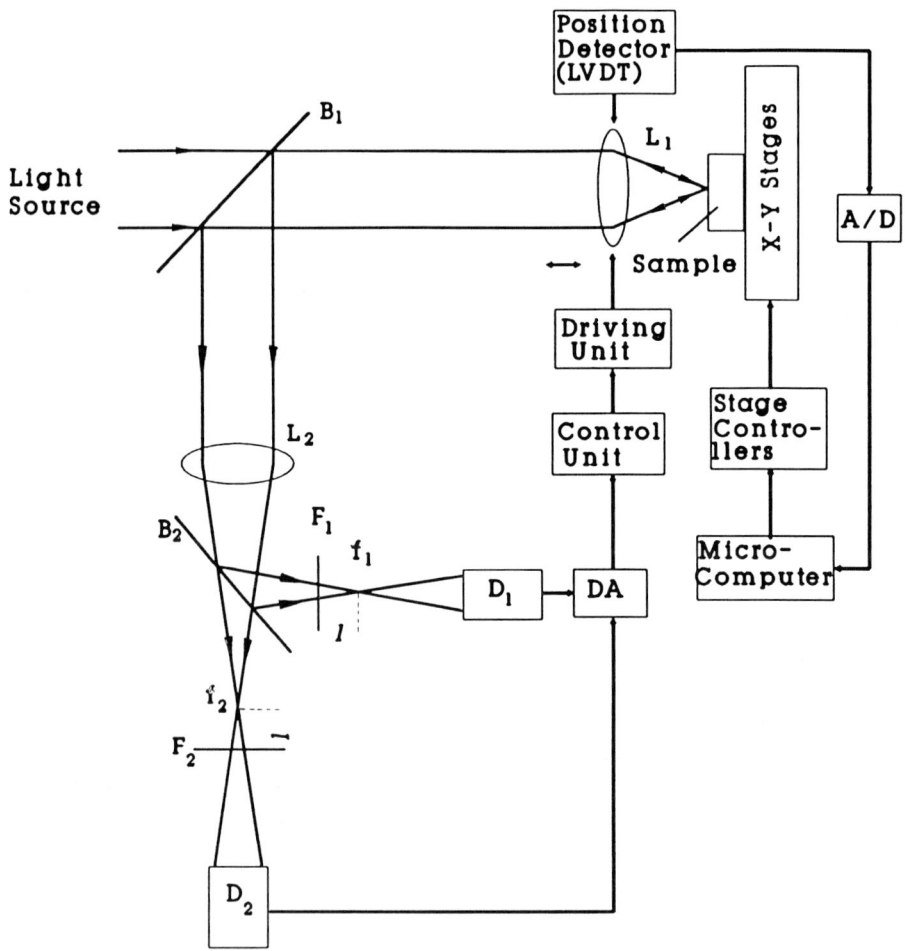

Figure 1.9 Schematic diagram of a differential focus detection system

1.4.1.2 Differential Detection Method[45–47]

Fig. 1.9 illustrates the measurement principle of a differential focusing method. It differs from the above-mentioned intensity focusing method in that a beam splitter BS_2 is inserted behind L_2, and two identical spatial filters F_1 and F_2, two large photo detectors D_1 and D_2, and one differential amplifier DA are used. If the light spot is in focus on the surface, the two focal planes of L_2 are formed at f_1 and f_2. F_1 and F_2 are used to match the light intensity distribution, F_1 is located in front of the focus plane f_1 at a distance l, and F_2 is located behind the focus plane f_2 at the same distance l. If the focused light spot is on the surface, the two photo detectors D_1 and D_2 would detect the same intensity of the reflected light, because F_1 and F_2 have the same relative position with respect to the focus plane of L_2. In this case the output from the differential amplifier DA is zero. As the surface height changes, it will cause an effective shift in the focal planes f_1 and f_2, approaching one spatial filter and retreating from the other. As a result, the power detected by one detector will increase, while the other one will show an approximately identical decrease of power. Therefore the differential amplifier DA will give a very good indication of the amount and sign of the height displacement of the evaluated surface, and the signal is used to control the position of the objective L_1.

1.4.1.3 Critical Angle Method[49,52,86]

The principle of the critical angle method is shown in Fig. 1.10. If the assessed surface is in focus (at position B), the reflected polarized laser light will pass through the objective L_1 which converts the light into a parallel flux. The critical angle prism positioned after L_1 will totally reflect the flux beam at the critical angle, and thus the same levels of light intensity are incident on the two photo detectors. Consequently, the out-of-focus signal from the differential amplifier becomes zero. Alternatively, if the surface is out of focus, say close to the objective L_1 i.e. at position A, the light flux diverges slightly after passing through the objective. As a result, the light on the upper side of the optical axis o-o strikes the prism at an angle smaller than the critical angle and the condition of total reflection is lost. Hence, this causes a refraction of the light passing out of the prism. On the other hand, the light on the lower side of the optical axis o-o strikes the prism at an angle larger than the critical angle and is totally reflected. Thus the intensities of light incident on the two photodetectors are different, and an out-of-focus signal is obtained from the output of the differential amplifier DA. An opposite phenomenon occurs if the surface is at position C, far from the objective; then the differential amplifier will give an out-of-focus signal with the reverse sign.

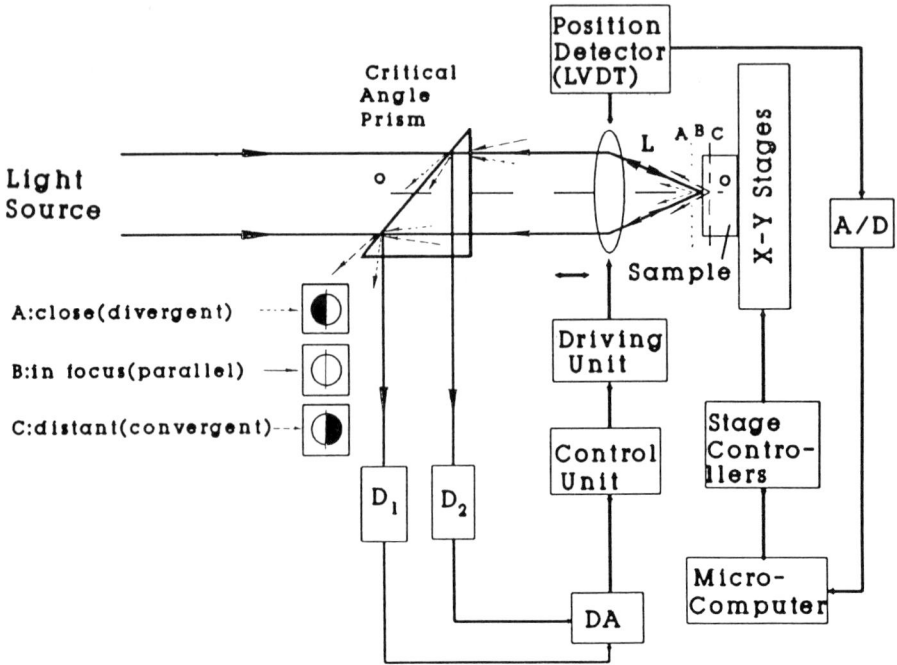

Figure 1.10 Schematic diagram of a critical angle focus detection system

1.4.1.4 Astigmatic Method[48,86]

The principle of the astigmatic method is shown in Fig. 1.11. The incident and reflected laser beams are separated by means of a polarisation beam splitter (PBS) and a quarterwave plate (QP). A cylindrical lens CL is inserted into the return path between PBS and the properly oriented quadrant diode QD. If the surface is in the focus of the objective L_1, a circular image is formed at the quadrant diode. Thus an 'in-focus' signal i.e. zero value is outputted from the operational circuit which carries out the mathematical operation:

$$E = \frac{(a + d) - (b + c)}{(a + d) + (c + d)} \tag{1-1}$$

where E is the focus error and the variables (a - d) are light intensities detected at different areas of the quadrant diode. If the tested surface is out-of-focus, an elliptical image will be formed at the quadrant diode. This usually occurs with the major axis lying in the plane of curvature of the cylindrical lens in the case where the specimen surface is too far from the objective (position C), or with the major axis perpendicular to that plane in the case where the specimen surface is close to the objective (position A). As a result, a focus error signal is outputted from the operational circuit.

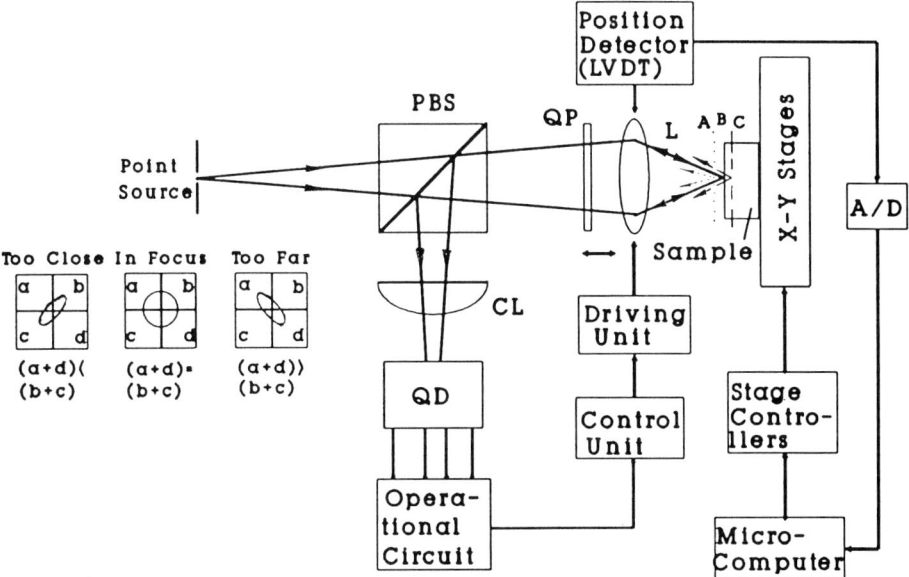

Figure 1.11 Schematic diagram of an astigmatic focus detection system

1.4.1.5 Foucault Method[5,40,43]

In the Foucault method shown in Fig. 1.12, a knife blade KB is inserted into the return path of the light from a narrow slit NS. An objective, L_2, is placed against the blade. With the aid of the knife blade, a circular image is formed at the objective, L_2, the nature of the image depending on which of the following three conditions is true; (i) If the tested surface is in focus (position B), a uniformly illuminated image (Fig. 1.12(b)) is obtained. (ii) If the surface is out-of-focus, say close to the objective L_1 (position A), the blade will interrupt part of the beam and the illumination of the image will be non-uniform as illustrated by Fig. 1.12(b), one side of the image becoming darker and the other lighter. The edges of the two areas are blurred and parallel to the direction of the edge of the blade. (iii) The positions of the darker and light areas on the objective L_2 are swapped with that of condition (ii), as the tested surface is far from the objective L_1 (position C). Since L_2 forms the image of the objective L_1 in the image plane I, two photodetectors D_1 and D_2, which are located at the image plane, would monitor changes in the brightness of the two halves of the image and output the defocused signal.

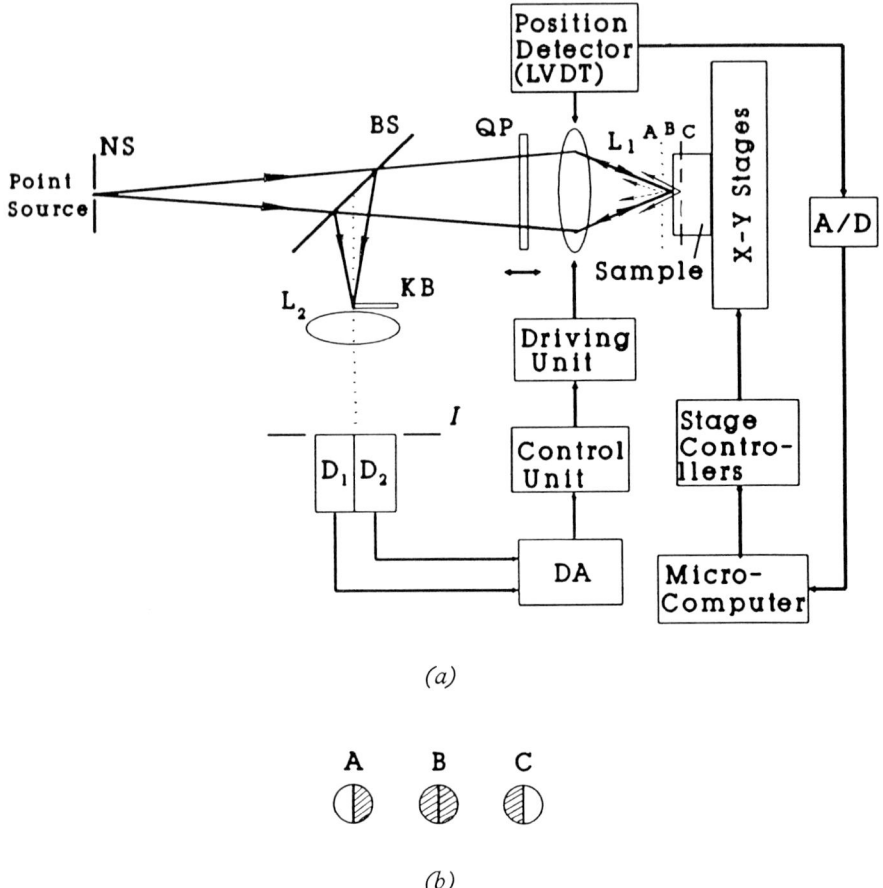

(a)

(b)

Figure 1.12 Schematic diagram of a Foucault focus detection system

1.4.1.6 Skew Beam Method[43,86]

This method uses an integrated slit-detector unit, D_1D_2, which is positioned in the plane of the object spot A (Fig. 1.13). An auxiliary narrow beam b_1 from the slit-detector passes the objective L off axis and the returning beam b_2 hits the detector D_1 or D_2. When the tested surface has an axial movement with respect to the objective, the surface height changes and leads to a defocusing Δz. As a result the photosignals detected by D_1 and D_2 are imbalanced, and the difference is a linear measure of the defocus around $z=0$.

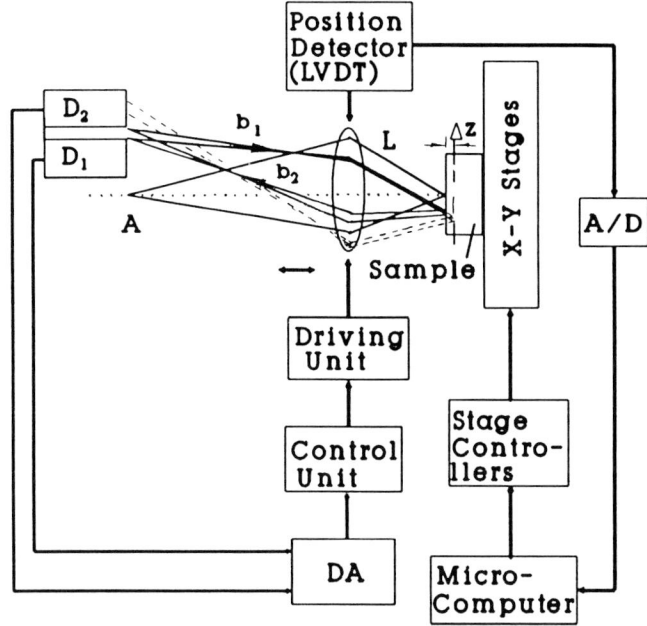

Figure 1.13 Schematic diagram of a skew beam focus detection system

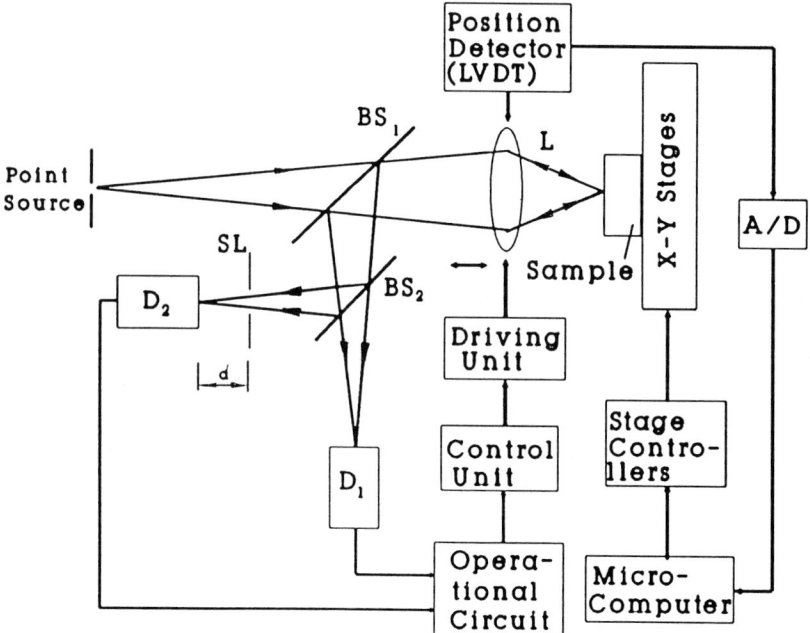

Figure 1.14 Schematic diagram of a defect-of-focus focus detection system

1.4.1.7 Defect-of-Focus Method[5,51]

Two photodetectors D_1 and D_2 are used in this method as shown in Fig. 1.14. The image point focused in the tested surface is used as the exploration spot. It is recognised that the light intensity distribution across the spot varies with the surface height and that the spot size variation is proportional to the defect of focus. By using a slit SL, which is mounted in front of D_2, the photosignals detected by D_1 and D_2 are normalised to cancel the local variation of the specimen reflection coefficient. As a consequence, the variation of intensity I_2 on D_2 depends only on the variation in the defect of focus, and can be expressed as a linear function of the defect of focus z:

$$I_2(z) = C \cdot Z + I_2(0) \ , \ C = \frac{2\Delta l(b-d)I_1(0)M^2}{\pi \Gamma d^2} \qquad (1\text{--}2, 1\text{--}3)$$

where $I_2(0)$ is the focus intensity detected by D_2 at $z=0$, and C is a constant, dependent on the pupil radius Γ, the magnification M of the objective L, the value b which is the distance from the objective to the image plane, the width Δl of the slit, the value d which is the distance between the slit SL and the photodetector D_2, and the in-focus intensity, $I_1(0)$, detected by D_1.

1.4.1.8 Confocal Method[159–176]

This is a somewhat different method from those mentioned above. It depends on eliminating scattered, reflected, or fluorescent light from out-of-focus planes while keeping the light in focus as strong as possible. This is achieved by inserting two pinholes P_1 and P_2 whose image are in focus with the specimen (Fig. 1.7(b)). When the specimen is located in the focal plane and the reflected light is focused on the pinhole, P_2, whose size can be as small as 0.2 μm, a strong signal is detected by the photodetector D. On the other hand, when the specimen is out of the focal plane, a defocused spot is formed at the pinhole P_2, and the measured intensity is greatly reduced, because the scattered, reflected or fluorescent light from out of focus planes will be strongly discriminated against by the pinhole P_2. Thus the term *confocal* here relates to the fact that the image of the illuminating pinhole and the back-projection of the detection pinhole have a common focus in the specimen. Due to the ability to produce strong signal within the depth of focus (normally 1–2 μm), this method is best suited for measuring a section of a sample tomography, hence *confocal scanning laser microscopy* (CLSM) is widely used in biological science.

 To realise 3-D measurement, two approaches are adopted. The first one[50] is the conventional method whereby surface topography is obtained by scanning a profile in the x-z plane based on the fact that focus is determined by finding the highest brightness and/or contrast; then a scan in x-y plane is realised. The second approach is based on the optical sectioning property: the capability for imaging a plane deep inside a specimen, without appreciable interference from the intervening and deeper lying layers. A series of optical sections of a specimen surface are recorded by scanning along the x-y plane with different step heights in the z direction. This stack of consecutive sections is then used to reconstruct 3-D information with the help of a computer. Since each section image is actually a volume element (*voxel*) and is in the form of an array, large amounts of data are recorded as the number of steps in the z direction increases. Therefore, data compression techniques have to be adopted.

A data compression technique was suggested by McCormick.[160] The first frame, which is the lowest, is stored in the first byte of a reference image memory. Subsequently grabbed frames are then compared pixel by pixel with the data stored in this reference image memory. If the intensity of the pixel in the most recently grabbed image is greater than the corresponding pixel in the reference image memory then the value of the grabbed pixel is stored in the corresponding first byte of reference image memory and the section number of the frame is stored in the next byte. This process is repeated for the number of sections required.

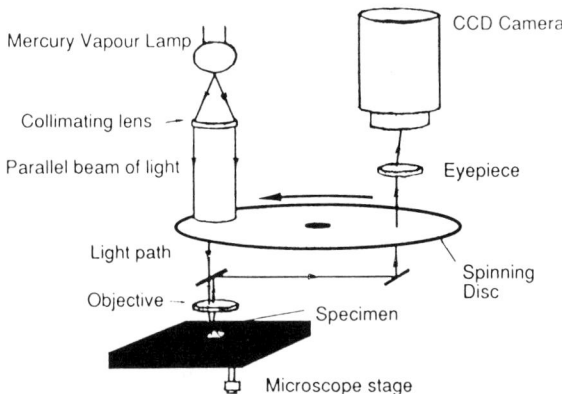

Figure 1.15 Schematic diagram of the Tandem Scanning Microscope (from McCormick)[160]

The scanning process used in confocal instruments can be either implemented by mechanical scanning in X-Y-Z directions as with other focus detection instruments, or to achieve higher scanning speed, by light scanning.[173] Conventional mechanical scanning can realise large scan ranges but with a relatively lower scan speed, whereas light scanning has a higher scanning speed. However, in biological research, measurement speed implemented by the series scan either mechanically or optically is not satisfactory for observing the tomographic change of living cells, since a real time 3-D image is required. A *tandem scanning microscope* (TSM) which uses a novel mechanical scanning system has been developed by Petran and Hadravsky.[176] With the help of a modern computer and image processing techniques it is able to produce real time images, a hence four dimensional measurement (X,Y,Z,t) can be realised. A schematic diagram of the TSM is shown in Fig. 1.15.

The microscope uses a Nipkow disc to provide the pinholes required for confocal imaging. The Nipkow disc is a perforated spinning silicon disc with thousands of square holes 50 micrometers in size arranged in Archimedian spirals. The pinholes are positioned so that, for each pinhole that supplies the illumination, there is an imaging pinhole diametrically opposite. As the disc spins at 1200 rpm these holes move across the whole field of view providing a full real time image. The advantage of using a spinning Nipkow disc is that at any moment many pairs of pinholes are being used to produce the image and so the object is being scanned in parallel trace fashion.

1.4.1.9 Properties of Focus Detection Methods

In these focus detection methods, although the surface height can be determined by directly detecting the light intensity, which is proportional to the defocus of tested specimen, a common detection method is to collect a defocusing signal which may be linearly proportional or non-linearly proportional to the defocus of the specimen. Then the defocusing signal is used to control a vertical driving device, which may be either a piezo-electric or a mechanical component, to reposition the specimen in focus. Surface height is sensed independently of the focus feedback control system, so as to maintain a linear measurement of surface height throughout the entire measurement range and a high scanning speed. Exceptionally, the confocal method reconstructs 3-D surface topography with a series of optical sections.

As for the scanning mechanism of the focus detection methods, both lateral and vertical scanning can be accomplished either by moving the specimen (*stage scanning*) or the light (*beam scanning*). The former can produce larger scan areas and the latter can attain higher scan frequencies. For many instruments used in engineering,[35–38] stage scanning is adopted in lateral scanning and the vertical scanning is implemented by moving the objective vertically. Movement of the objective is realised either by a linear/step/DC motor or by a piezo-electric driving device. Motor driving is suitable for measuring a variety of rough surfaces, whilst piezo-electric driving is suitable for measuring very fine surfaces and for fast scanning.

The lateral resolution of focus detection methods is related to the size of the light spot and the resolution of the scanning driving device. The vertical resolution depends on the particular focus detection method adopted. Table 1.1 summarises some characteristics of the various focus detection methods. It is seen that the lateral resolution of all methods and commercial instruments are within 0.1–2 μm, and that differential, critical angle and astigmatic methods possess nano or sub-nanometer resolution in the vertical direction. Since the introduction of feedback control in measurement systems, the vertical measurement range of this type of optical system is no longer limited by the wavelength of the incident light, which is a critical limitation for other kinds of optical systems such as interferometric and light scattering systems,[3,90] The maximum range can be up to several millimeters, which is similar to, or greater than, the range of the conventional stylus instrument. The measurement speed is also very high and, for most methods, scanning on a surface with large amounts of data (256x256) can be accomplished within a minute, and real time scans can even be realised by using the TSM instrument.[160]

Although focus detection instruments have great advantages over stylus instruments because of their non-contact mode of operation, there are some circumstances where the focus detection instrument is inferior to the stylus instrument. These are:

- Firstly, the focus instrument is more sensitive to surface inclination. If the surface inclination exceeds a critical angle, depending among other things on the reflective behaviour of the sample, the reflected light misses the objective lens (Fig. 1.7). The more the surface is inclined, the more difficult the focusing becomes. For surfaces with an ideal reflection the critical angle lies between 10° and 15°.

- Secondly, the reflectivity of samples is a crucial factor in influencing measurement values. Focusing can become impossible if the surface to be measured reflects less than 4% of the radiated light. The instrument is more sensitive to surface inclination with low reflectivity of the sample. In addition, deviations in the measurement are also possible with components having greater reflectivity if the optical contrast of the surface is very high, e.g. on sharp colour of light/dark transitions.

- Thirdly, measurement results are influenced by the microgeometry of the sample. Distorted results may be obtained from measuring a spatially small tip, or a pore, or a sharp step feature.

- Fourthly, as with other optical systems, focus detection systems react to impurities of any kind; the sample surface to be measured must be free from dust, loose particles and liquid (water, oil) to ensure reliable results. This restricts their application in workshop practice.

Table 1.1 Some characteristics of focus detection systems

	Vertical Resolution (μm)	Horizontal Resolution (μm)	Vertical Range (μm)	Measurement Speed/Time
Intensity[5,42]	0.1	0.5	50	
Differential[45–47]	0.002	<2	>1000	0.2–1 mm/s
Critical Angle[49,52, 86]	0.0002	0.65	3	10 mm/s and <5 min. for 1000x1000 in mm^2
Astigmatic[48,86]	0.002		4	
Foucault[5,40,43]	0.01	<1	60	
Defect of Focus[5,51]	<<0.1	2	>20	
Confocal[159–176]	0.1	0.1	380	5s for 256x256 and 20s for 1024x1024
TSM[160]	0.02	0.2	50	Real time
Zeiss LSM[38]	<0.05	0.25	>100	0.5–64s for 512x512
Rodenstock RM 600[35]	0.002	1	600	28s for 30,000 points in 2x2 mm^2
Perthen Focodyn[36]		1	500	0.1–0.5 mm/s
UBM UB16[37]	0.005	1.5	500	

1.4.2 Interferometric Instruments

Optical interferometry has, for a number of decades,[60] been used to measure surface topography and many interferometric instruments based on two or multibeam interferometers have been proposed.[60,84,85] However, since interferometric images or interferograms which show optical fringes with patterns of light and dark bands are very difficult to interpret and translate into useful measurements of surface texture, this technique was limited to qualitative analysis and visualisation of surface topography until the middle of the 1970s. It was the development of the modern computer, optical-electronic and sophisticated computer graphics techniques that made profiling interferometry possible. Now optical interferometry has become one of the most popular profiling techniques used to measure 2-D and 3-D surface topography. In some applications, such as in profiling optical components and magnetic tapes, the interferometric instruments have become routinely used.[66–69]

The underlying principle of interferometry is that two light waves, when brought together, interfere with each other. If the crest of one wave coincides with the trough of the other, the interference is destructive and the waves cancel out. On the other hand, if two crests or two troughs coincide, the waves reinforce each other. Then, an optical fringe with parallel and dark bands similar to that in Fig. 1.16 would be produced.

Figure 1.16 An example of optical fringe

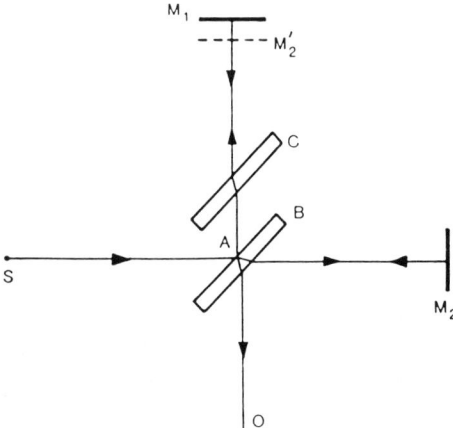

Figure 1.17 Schematic diagram of the Michelson interferometer

A typical interferometer, the Michelson interferometer, is shown Fig. 1.17. In the interferometer an optical wavefront from a source, S, is split into two beams with coherence and nearly equal amplitudes by a beam splitter B. These two beams are reflected back at a smooth flat reference surface M_1 and a specimen surface M_2, and return to B where they are recombined and emerge at O. Texture on the specimen's surface changes the distance travelled by the second beam. When the beams recombine, some parts of the second beam will be in phase with the first, while others will be out of phase. The spatial relation between the two beams contains detailed information about the topography of the surface. In fact, if an ideally flat surface is measured with an interferometer, the fringes in the obtained interferograms would be straight-lined and equally spaced from each other. If a practical surface which is not ideally flat is measured, the fringes will depart from being straight and equally-spaced. The undulations of each fringe reveal the peaks and valleys of the profile of the tested surface. Therefore, the aim of interferometric instruments for 3-D surface topography is to interpret the fringes and transform them to produce the 3-D surface topography.

All currently commercially available interferometry instruments for 3-D surface topography measurement are actually a combination of an interferometer, a microscope and a microcomputer. A large number of interferometers, such as the Michelson, Fizeau, Mirau, Linnik and Nomarski interferometers can be used to construct 3-D interferometric instruments. According to profiling mechanisms, these interferometers can be classified into two major categories, i.e.

- interferometers e.g. Michelson, Fizeau, Mirau and Linnik; these measure surface topography height directly.
- interferometers e.g. Nomarski; these measure the slope of the surface.

The former interferometers have the obvious advantage in getting surface height directly; however, they are very sensitive to mechanical vibration, air turbulence and temperature variation. The latter types have the advantage of being sensitive to surface height variation, insensitive to flatness error of a mechanical scanning driving device and less affected by environmental vibration. Since surface height is obtained by the integration of the slopes for slope measurement instruments, a digital integration error may be introduced. On the other hand, measurement direction is essential for slope measurement interferometers.

In recent years, some interferometric techniques such as phase shifting,[61–65] heterodyne,[57–59] common-path polarisation,[56] differential interference contrast (DIC)[72,73] and scanning differential interferometry[74–75] have been successfully applied in construct profiling interferometric instruments. Two of them, phase shifting interferometry and scanning differential interferometry are now commonly used for 3-D surface topography measurement and commercial instruments of these types are available.

1.4.2.1 Phase Shifting Interferometric Instrument

The phase shifting interferometry technique used for surface topography measurement was developed by Bruning[61] in 1974. Surface height is determined not by looking at the interference fringes and by measuring how they depart from being straight and equally spaced, but by finding the phases of the interference pattern produced by the two reflected wavefronts from the reference surface and the specimen surface at all co-ordinate pairs (x,y). Once the phase, $\phi(x,y)$, is obtained across the interference field, the corresponding height distribution h(x,y) is determined by the equation[89]

$$h(x,y) = \frac{\lambda}{4\pi}\Phi(x,y) \tag{1-4}$$

where λ is the wavelength of the incident light. Obviously, the key purpose of the phase shifting interferometric technique is to determine the phase $\phi(x,y)$. This is achieved by measuring three or more interference patterns, each associated with a slightly different axial position of the reference or the specimen surface. In this case, the intensity of each interference pattern at position (x,y) is a function of the initial phase $\phi(x,y)$ and a shifted phase α_i (i=1,2,...,n). That is

$$I_i(x,y) = A + B\cos[\Phi(x,y) + \alpha_i] \quad (i + 1,2,...,n) \tag{1-5}$$

where A is the average intensity, B is a constant proportional to A. $B\cos[\Phi(x,y)+\alpha_i]$ represents the interference term. According to synchronous detection techniques in communication theory,[89] if the shifted phase α_i is an increment by $i(2\pi/n)$ then there is a relationship given by

$$\tan\Phi(x,y) = -\frac{\sum\limits_{i=1}^{i=n} I_i(x,y)\sin[2\pi(i-1)/n]}{\sum\limits_{i=1}^{i=n} I_i(x,y)\cos[2\pi(i-1)/n]} \tag{1-6}$$

taking n=4 as an example,

$$\Phi(x,y) = \tan^{-1}\frac{I_4(x,y) - I_2(x,y)}{I_1(x,y) - I_3(x,y)} \tag{1-7}$$

is obtained.

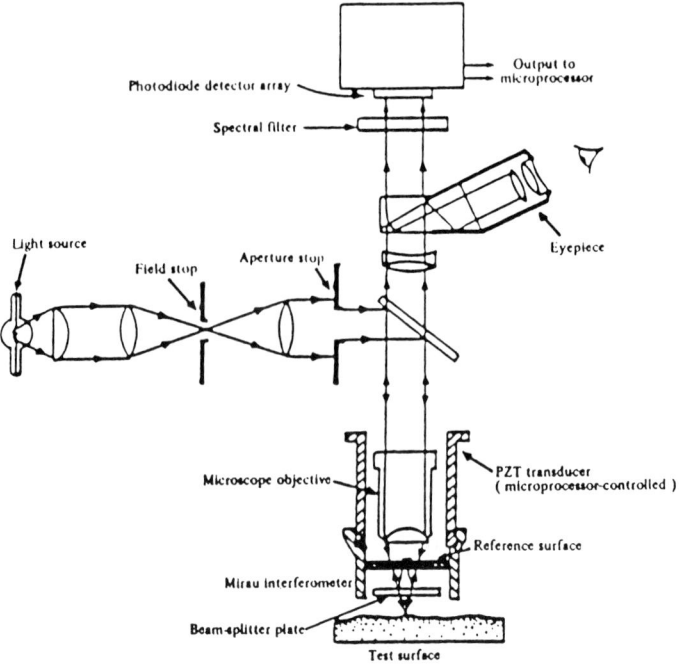

Figure 1.18 Schematic diagram of a phase shifting interferometry instrument (from Bhushan)[64]

In order to realise phase shifting, to detect the phase of the interference patterns and to implement the mathematical algorithm, a driving device is used for moving the reference surface or the specimen. An area photodetector array and a microcomputer are needed in the measurement system. Fig. 1.18 shows a schematic diagram of such an interferometric system. An interferometer is located at the bottom of the instrument, and the reference surface is attached to a piezoelectric transducer (PZT). The phase shifting is realised by exerting a voltage which is from a digital to analogue (D/A) converter driven by a microcomputer to the PZT, which then produces a force to drive the reference surface through a displacement. In order to minimise the influence of vibration induced due to stepping the phase shift, the integrating-bucket technique[64,67] for moving the reference surface is used preference to the step moving technique. As such, the phase α_i is modified to $\alpha(t)$, and it changes at a constant rate. An optical microscope objective above the interferometer magnifies the interference pattern produced by the interferometer, and then the magnified interference patterns are detected by an area image sensor e.g. *charge-injection-device* (CID). The analogue video signal from the image sensor is then digitised by a video frame grabber. Since the phase shifting $\alpha(t)$ is designed to have a constant velocity and to be synchronised with the image grabbing, each time the CID area array is read out because the total phase of the interference term has changed by 90 degrees compared with the previous phase. This results in a stable image digitising process. The digital image is stored in the frame grabber's memory. Until three or more (dependent on the software algorithm) images are captured, a two-dimensional array of height elements of the surface is calculated by the microcomputer. The array data can be further processed by the powerful (fast and large capacity of memory) microcomputer to obtain a variety of 3-D surface topography visualisation pictures and 2-D and 3-D surface roughness parameters. All the measurement and analysis processes are under the control of the powerful microcomputer.

1.4.2.2 Scanning Differential Interferometric Instrument

Differential interferometric instruments are normally based on a Nomarski microscope which can obtain interference patterns of two light beams reflected from two neighbouring incident spots. Usually, the Nomarski microscope is used for the qualitative assessment of surface topography, and it makes it easy for scratches, dust particles, cleaning marks, fingerprints, dig, small bumps, microstructure, grain structure, and all kinds of machining marks on moderate to highly reflecting surfaces to be seen. In 1979, Lessor et al.[72] developed a method to obtain quantitative surface topography with the Nomarski microscope. Since it is based on quantizing the *differential interference contrast image* (DIC), the technique was not as efficient as the later developed *scanning differential interferometric instrument*[74] for the measurement of 3-D surface topography.

One of the most successful scanning differential interferometric instruments was developed by Bristow.[74] Contrary to the DIC method, this instrument is based on the fact that the phase difference of two neighbouring beams focused on the surface is related to the surface height difference, and hence the difference in strength of the two beams is proportional to the surface height difference. A schematic diagram of such a system is shown in Fig. 1.19(a). The system includes two parts. One is the light illuminating and phase detecting part, which is located at the left hand side of Fig. 1.19(a). The major components of this part are a laser head, a non-polarising beam splitter (the upper one), a polarising beam splitter (the bottom one) and two photodetectors. The second part is the interferometer, which is located at the right hand side of the figure, and is placed on a translation stage. The major components of this part are a Nomarski (or modified Wollaston) prism, an objective and two mirrors in a penta prism arrangement. The collimated laser beam passes to the mirrors and is reflect to the Nomarski prism. This prism is composed of two wedges of birefringent material that are cut and assembled in such a

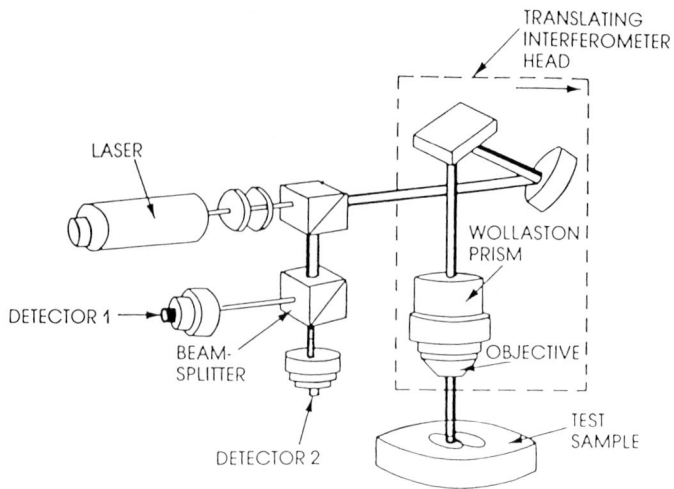

(a) Construction of the instrument (from Bristow)[74]

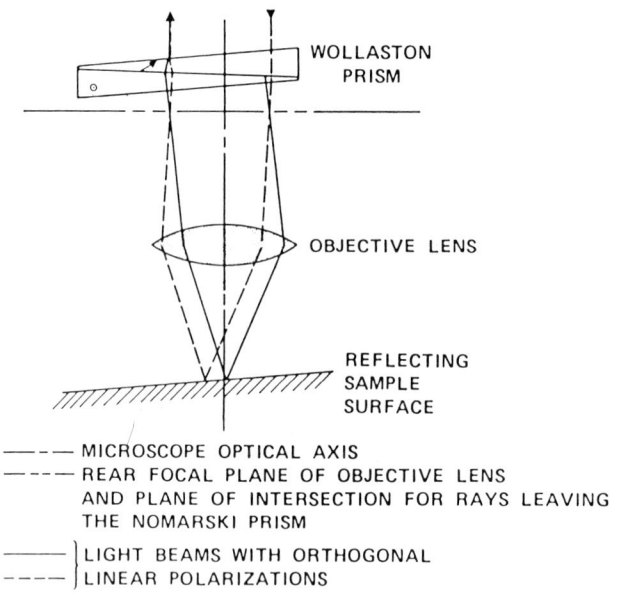

----- MICROSCOPE OPTICAL AXIS
----- REAR FOCAL PLANE OF OBJECTIVE LENS
        AND PLANE OF INTERSECTION FOR RAYS LEAVING
        THE NOMARSKI PRISM

——— | LIGHT BEAMS WITH ORTHOGONAL
----- | LINEAR POLARIZATIONS

(b) Profiling principle of the Nomarski prism (From Lessor)[73]

Figure 1.19 Schematic diagram of a scanning differential interferometry instrument

manner as to split an incoming beam into two orthogonal polarized components as shown in Fig. 1.19(b). The polarized beams are then focused on the specimen surface with a focal spot diameter of about 1–1.6 μm, and a separation of about $1/4$ of the diameter. The beams reflected from the specimen surface will spatially recombine at the Nomarski prism and retain their polarisation identities until they pass through the non-polarising beam splitter. The collinear polarized beams are finally split into their respective components by the polarising beam splitter and directed to either one of the photodetectors. The signal difference of the two photodetectors is proportional to the height difference or the surface slope between the average surface levels in the areas illuminated by the two beams. As the translation stage moves, the second part of the system will cause the polarized focal spots to scan across the surface. Hence a series of surface slope data along the translation direction is obtained, and a profile is then calculated by integrating the slope data at each point. The 3-D surface topography is obtained finally by adding one more translation direction which is perpendicular to the previous one as in 3-D stylus systems.

1.4.2.3 Properties of Interferometric Instruments

The main advantage of interferometric instruments over other optical instruments is that they have sub-ångström resolution in the vertical direction. They can therefore be used to measure the topography of very fine surfaces such as optical lenses, laser gyro mirrors, diamond turned optics, floppy disks and magnetic heads whose reflectivities are above 1%. However, their vertical measurement ranges are limited by the wavelength of the illuminating light. Although some measures can be introduced to extend the measurement range,[3,55] the measurement range of many commercial instruments[76–79] is still confined to tens of microns.

The lateral resolution and range of phase shifting interferometric instruments are related to the magnification of the objective used and the pixel spaces of the area CID array. Normally the pixel spacing is about 25 μm, so if a 10x objective is used, the lateral resolution is 2.5 μm and the total measurement area is 2.5(m-1)x2.5(n-1), where m and n are dimensions of the area CID array. Compared with the phase shifting interferometric instrument, a scanning differential interferometric instrument is able to analyse a larger measurement area due to the scanning being driven by a mechanical device. This area can be up to 100x25mm which is similar to that of stylus instruments. However, one difference with the stylus instrument is that an autofocus attachment is used in the scanning differential interferometric instrument. While making long scans over a curved surface, the system will dynamically adjust the focus, point-by-point. Thus it enables measurements of both roughness and waviness simultaneously, and is suitable for measuring a spherical and cylindrical surfaces. The lateral resolution depends on the resolution of the objective and the translation stage, the aberrations in the optical system, the separation of the two beams focused on the surface, and the actual surface topography measured. Usually it is in between 0.8–1.6 µm. Table 1.2 summarises some main characteristics of four commercially available interferometric instruments. The first three are phase shifting instruments and the fourth is the scanning differential instrument.

Table 1.2 Characteristics of Three Commercially Available Interferometry Instruments Vertical

	Vertical Resolution (nm)	Lateral Resolution (µm)	Vertical Range (µm)	Maximum Mea. Area (mm x mm)	RMS Repeatability (nm)	Measurement Time
Wyko Topo-3D[76]	<0.1	0.35–17.4	15	4.44x4.44	0.003	465 ms
Zygo Maxim 3D[77]	0.05	0.36–21.4	40	6.74x6.74	<0.1	<5 s
Micromap Promap-512[78]	0.02	0.55–8.8		2.7x2.5	<0.1	
Chapman MP2000[79]	<0.1	0.5–100	15	100x25	0.1	2 mm/s

Beyond the considerations discussed above, there are two problems associated with the interferometric technique.

- One is that an interferometer requires that the tested sample has an optical constant that does not vary over the area being mapped, since a change in the optical constant changes the phase difference of the interferogram, which in turn influences the height difference.
- The other problem is that the influence of the reflecting coefficient of the sample surface must be taken into consideration. Surfaces of low reflectivity, e.g. plastic or glass, necessitate a correspondingly small reflection coefficient of the reference surface. Therefore, a change of the reference surface may be needed for measuring a surface with a range of different reflecting coefficients.

1.5 NON-OPTICAL SCANNING MICROSCOPY

Some optical scanning instruments or microscopes have been described in the last section, in which some real optical lenses are involved. However, there are some scanning microscopes which do not have real optical lenses, and images are not observed through optical lens, but from a computer image monitor or a cathode ray tube (CRT). There are two types of non-optical scanning microscopes. The first type is the electron microscope and includes the *scanning electron microscope* (SEM), the *transmission electron microscope* (TEM) and the *scanning transmission electron microscope* (STEM). The other type is the *scanning probe microscope* which includes the *scanning tunnelling microscope* (STM), the *atomic force microscope* (AFM), the *scanning field-emission microscope* i.e. the *topografiner*, the *scanning capacitance microscope* (SCM),[216–218] the *scanning thermal microscope,*[223–224] and the *scanning acoustic microscope* (SAM).[225–228] In principle, the scanning probe microscope is similar to the stylus instrument in the tip and scanning manner except that a different probe mechanism is used instead of the contact mechanical stylus probe. So almost all the scanning probe microscopes were originally designed to have the ability for the quantitative measurement of 3-D surface topography. Physically, the scanning process of the electron microscope is different from the scanning probe microscope. Only two dimensional images can be obtained by the conventional electron microscope. In the last two decades, some stereology techniques have been developed to obtain quantitative 3-D information using electron microscopy. Non-optical scanning microscopy has been popular in many disciplines, including materials, biology, chemistry, medicine and manufacture, for the quantitative analysis of 3-D topography of very fine surfaces. Especially as a result of the advent of the STM and the AFM, the atomic structure of surfaces of different materials or biological cells can be identified. The rest of this section is devoted to a review of some important and commercialised non-optical scanning microscopes, SEMs, STMs and AFMs.

1.5.1 Electron Microscopy

The electron microscope was originated from the invention of the first TEM by Knoll and Ruska[13] in 1931. It was realised that an electron microscope could be made by focusing a scanning electron beam onto a specimen surface and then recording the emitted current as a function of position. Subsequent research resulted in the emergence of the first SEM.[15] However, it was thirty years later that the first commercial SEM was launched by the Cambridge Instrument Company[16–17] in 1965. The basic structure of a SEM as used in its most common mode, the emissive mode, is shown schematically in Fig. 1.20.[34] There are three groups of components. The first group is the electron optical column which consists of an electron gun, two or three magnetic lenses and two sets of scanning coils. As an electron beam from the electron source flows through the lenses, these lenses cause the beam to be focused onto the specimen surface. The scanning coils placed in front of the final lens cause the electron spot to be scanned across the specimen surface in the form of a square raster in much the same manner as a television screen. The currents passing through the scanning coils are made to pass through the corresponding deflection coils of a CRT so as to produce a similar but larger raster on the viewing screen in a synchronous fashion.

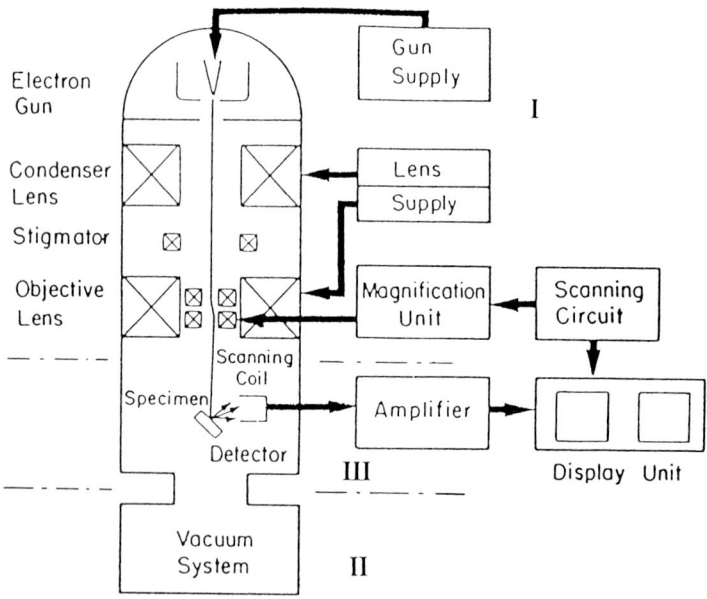

Figure 1.20 Schematic diagram of the SEM (from Johari) [34]

The second group of components is a vacuum system which includes the specimen chamber and stage. A pumping system provides the necessary vacuum both in the electron optical column and in the specimen chamber. The stage allows the specimen to be examined at the required angle relative to the incident electron beam. According to electron emission theory, several phenomena would be produced as the incident electron beam (the primary beam) bombards the specimen surface. These phenomena are:

1. Emissive (secondary electrons).

2. Reflective (backscattered electrons).

3. Absorptive (leakage current).

4. Transmission.

5. Beam-induced conductivity.

6. Cathodoluminescence.

7. X-ray.

8. Auger electron.

9. Radiation damage.

Any one of the phenomena can be detected by using a proper detector and used to form an image on the CRT screen. Thus the signal detection, amplifier and image display units consist of the third group of components. Usually, the secondary electron effect is most commonly used in commercial SEM. The emitted secondary electrons from the specimen surface strike the detector and the resulting current is amplified and used to modulate the brightness of the CRT screen.

The great advantage of the conventional SEM is that it has very high spatial resolution (several nanometers) and a much larger depth of field than the light microscope. However, since the conventional SEM only provides a 2-D topography image of a surface, it merely affords a qualitative interpretation of the surface; quantitative data about the height of surface features is not readily obtained. To overcome this, some quantitative techniques[30–34,177–189] have been developed since the end of 1969's. These techniques can be classified into two categories. One is based on stereopair,[177–184] the other is based on the fact that the backscattered electron signal is proportional to the inclination of the surface along the scanning direction.[185–189] Both of them are described below.

1.5.1.1 Stereopair Technique for Quantizing SEM Images

This technique is used for obtaining 3-D information from the secondary electron images. It is clear that the only prerequisite for a true 3-D interpretation of a scene is that the images presented to the observer's two eyes should contain parallax. This is the difference between the distance apart of two points in the two 2-D images forming the stereopair, measured along the line perpendicular to the tilt axis between them. In order to obtain stereo information, two SEM images with different tilt angles must be produced. This can be implemented by either tilting the specimen or the incident electron beam between two exposures. The difference between the two tilt angles is usually set at 6–10 degrees. According to the Nankivell algorithm[177,178] the third dimension of the specimen topography can be calculated from the two tilted SEM images. An example of the algorithm is shown in Fig. 1.21. The projection of the object $P_1P_0P_2P_3$ on to the horizontal plane is a straight line $PP_1^0P_2^0(P_3^0)=\Delta S$ (Fig. 1.21(b)), the point P_3^0 is overlapped with the point P_2^0, hence the object height ΔH is unable to be distinguished. If the object is tilted $\theta/2$ and $-\theta/2$ around the tilt axis P_0, the two projections of the tilted object on to the horizontal plane would be as shown in Fig. 1.21(a) and Fig. 1.21(c). Let $P_1^+P_3^+=\Delta S^+$ and $P_1^-P_3^-=\Delta S^-$. The parallax of the two tilted projection images of the point P_3 is $\Delta P=\Delta S^- - \Delta S^+$. Where

$$\Delta S^+ = \Delta S\cos\frac{\theta}{2} - \Delta H\sin\frac{\theta}{2} \;,\; \Delta S^- = \Delta S\cos\frac{\theta}{2}+\Delta H\sin\frac{\theta}{2} \qquad (1\text{--}8)$$

If the magnification of the image system is M, the object height ΔH can be obtained from the parallax ΔP. That is

$$\Delta H = \frac{\Delta P}{2M \sin \dfrac{\theta}{2}} \qquad (1\text{--}9)$$

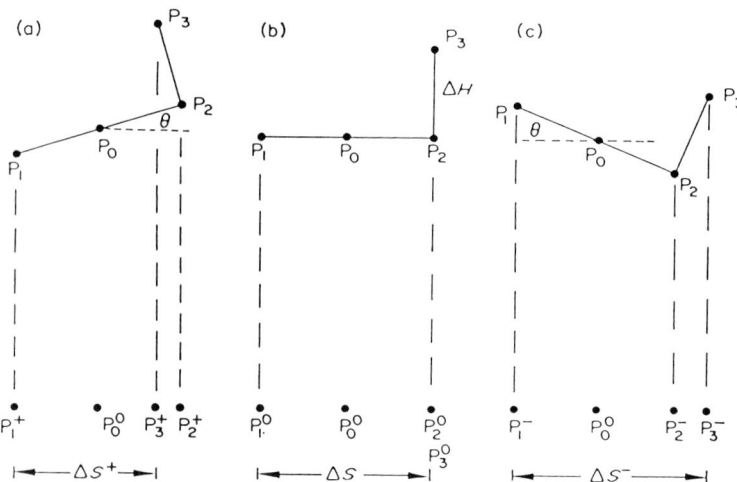

Figure 1.21 The derivation of the standard parallax equation (from Hudson)[178]

As an example, suppose the tilted angle θ is 10 degrees, the parallax ΔP is 0.1mm with the magnification M equal to 20000x, then the object height is about 28.6nm. It is obvious from the formula that the accuracy of stereo calculations depends on the accuracy of parallax measurement and on the accuracy of the tilt angle and the magnification.

After stereopair images are recorded with an SEM, a stereo-compara-tor[180] is used to measure parallaxes, i.e. the difference in the projection distance between corresponding pairs of points in the two images. The 3-D co-ordinates of relevant surface features are derived by the instrument, and it then outputs data directly to a microcomputer for further digital image and parametric analysis.

1.5.1.2 Direct Integration for Quantizing SEM images

Instead of using secondary electrons in the stereopair technique, backscat-tered electrons are detected in the direct integration method. With this method, a surface profile is obtained by the integration of the backscattered electron signal, the signal being in proportion to the surface inclination along the scanning direction. A schematic of this type of quantitative SEM system is shown in Fig. 1.22.[186] It consists of a conventional SEM, a minicomputer, an A/D converter, a clock pulse oscillator and some other auxiliary peripheral devices, and aims to take the signal of an indicated frame of the image on the CRT of the SEM into the computer and to carry out its image processing. The backscattered electron image signal is detected by a detector and transmitted to both the CRT of the SEM for 2-D image display and to the A/D converter for digitising. The synchronised sample pulse is obtained from the clock pulse oscillator which is controlled by the scanning synchronised signals and the start pulse output from the computer. The computer manipu-

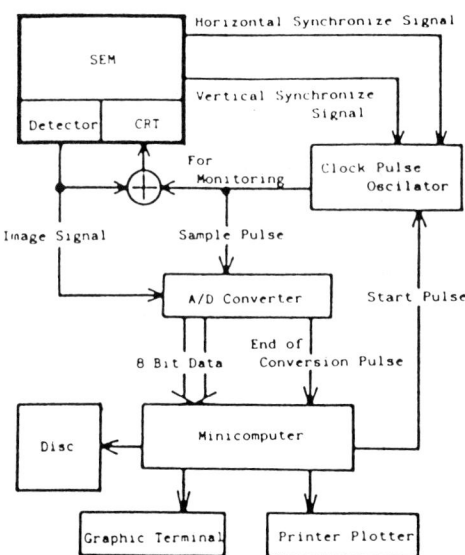

Figure 1.22 Block diagram of a direct integration SEM system (from Sato)[156]

lates the data sampling process and carries out the post-processing of the digitised SEM image. The peripheral devices are used to store the digitised data and to display or hardcopy the digitised 3-D information.

1.5.1.3 Properties of Electron Microscopy

Compared with the above mentioned two quantitative techniques, the direct integration technique is more intuitional than the stereopair technique. In addition, it is able to achieve faster measurement speed. However, with the stereopair technique, better spatial resolution can be obtained due to the secondary electrons rather than the backscattered electrons being detected. In principle, the spatial resolution is dependent on the diameter of the electron beam, the current of the beam and the electron phenomenon detected. For current commercial SEMs, a spatial resolution of several nanometers is readily available.[190–192] Normally, the spatial resolution will not be greatly affected by either of the quantization techniques. Measurement range in the lateral direction depends on the magnification. A small magnification can have a larger measurement range, say up to several millimetres. The vertical resolution of the quantization techniques relies on several factors. For the stereopair technique, the accuracy of the measured parallax, the tilt angle and the magnification mainly dominate the maximum vertical resolution. The vertical resolution using the direct integration technique depends on the low and high frequency noise induced in the backscattered signal, and the accumulative integration error. Several tens of nanometer resolution[181,183] and nanometer resolution[185,186] in the vertical direction were claimed for both techniques. The vertical range of the quantization technique is a few hundreds of microns.

The quantitative SEM faces the same problems as an ordinary SEM. It takes time to prepare a sample; a non-conductive sample has to be coated before being measured, and it is difficult to measure very smooth surfaces since there is nothing to focus on. In addition, the speed of the stereopair technique is very slow; tens of minutes is required to record a frame of a 3-D topography image.

1.5.2 Scanning Probe Microscope

All scanning probe microscopes use the same measuring technique in that they measure of surface topography using a very sharp probe positioned in very close proximity to a surface. The probe scans the surface in a controlled fashion, and the topography is logged during the scanning process at regular intervals. The major difference between the different types of scanning probe microscopes is the probe mechanism which can be based on various physical phenomena. Field emission, scanning tunnelling current, magnetic field strength, thermal conductivity, capacitance, interatomic force, etc. are all valid for constructing different scanning probe microscopes. An earlier scanning probe microscope was reported by Young[29] in the early 1970s. This instrument adopted the field emission effect and was called *topografiner*. A vertical resolution of 3 nm and a spatial resolution of 400 nm were claimed by the author. For various reasons this instrument was never developed further, and therefore no commercial instrument is available.

The most successful scanning probe microscope for 3-D surface topography measurement is the scanning tunnelling microscope (STM) which was invented by Binnig and Rohrer[80–82] in 1981. It has a spatial resolution of about 1 Å and a vertical resolution close to 0.01 Å and this established the scale of magnitude of the STM in many academic research disciplines. This has advanced surface metrology from the micron scale of magnitude to the atomic. Just five years after the invention of the STM, Binnig and Rohrer were jointly awarded the 1986 Nobel prize for physics due to their outstanding contribution to the STM. Now the STM and subsequently developed AFM have played important roles in the research of nano- or sub-nanometer topography including atomic and molecular structures; they have been widely used in material, manufacturing, optical, chemical and biological engineering/science.[193–204] Not only are many commercial instruments available, but also many research groups are able to design STM and AFM for their own applications.[205–207,213] This is mainly due to the clearly demonstrated theory and the simplified structures of the STM and the AFM.

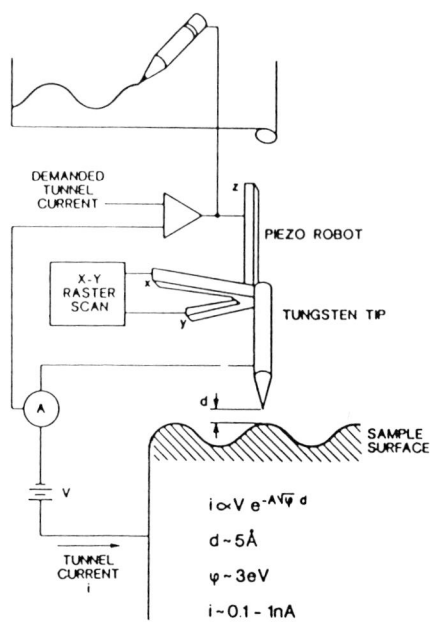

Figure 1.23 Principle of the STM (from Besenbacher)[196]

1.5.2.1 Scanning Tunnelling Microscope

The basic principle of the STM is straightforward (Fig. 1.23). An atomically sharp metal tip is mounted on a three dimensional micro-driving unit which is in tripod form (or a tube or a laminated slab) with three piezo-electric transducers (Fig. 1.23). The tip serves as one tunnel electrode and is brought to a conductive specimen surface which serves as the second tunnel electrode. The spatial separation is in the scale of several ångströms. According to quantum mechanics, electrons on a surface behave both as particles and as waves. As a result of this behaviour, the electrons on the surface of a material form an electron 'cloud' as represented in Fig. 1.24.[208] The gap between the metal tip and the conductive specimen presents a barrier to the electrons passing freely between them. However, if a bias voltage (2 mV – 2 V) is applied between the tip and the specimen, a small number of electrons, depending on the gap size, can tunnel through the gap and a tunnel current (on pA – nA scale) flows under ultra-high vacuum (UHV) condition. Since the electron cloud on a surface is less dense and the density shows an exponential decrease with increasing distance from the surface, the tunnel current increases exponentially as the gap is decreased. This implies that a small change in tip-to-sample distance produces a large change in tunnelling current, e.g. a change in distance of 1 ångström can cause the tunnelling current to vary by an order of magnitude or more, and this endows a remarkable vertical sensitivity to the STM.

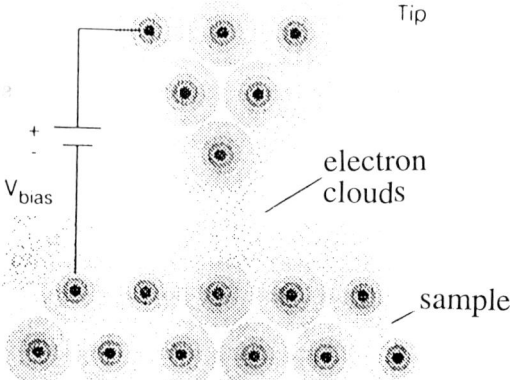

Figure 1.24 Electron clouds on surfaces of the tip and the sample (from Burleig Instruments) [208]

Based on the fundamental principle of the STM, the topographical height of the specimen surface can be obtained with two modes.

- *Constant tunnelling current mode* – The tripod drives the tip raster scanning across the surface, and a constant tunnelling current is maintained during the scanning process. In other words the distance between the tip and the specimen surface is kept constant during the scanning process. This is achieved by adopting a feedback control system which keeps the tunnelling current as well as the distance constant. When the device detects an increase in tunnelling current, it compensates by retracting the tip slightly from the surface. Conversely, when it detects a decrease in tunnelling current, it extends the tip closer to the surface. Thus topography heights at different positions for this mode are the variation of the voltage applied to the piezo-electric transducer in the Z direction which is necessary to maintain the constant current.

- *Varied tunnelling current mode* – In this mode absolute height of the tip is fixed, and the tunnelling current varies as the distance between the tip and the surface changes during the scanning process. The topography height of the surface is thus determined from the measured tunnelling current.

In most STM systems[205,206,213] the first mode i.e. the constant tunnelling current mode is adopted. This mode tends to provide a 'smoother' topography and is best for the measurement of an irregular surface. However, since the feedback control is constantly making height adjustment to the tip, the measurement speed is slower than the varied tunnelling current mode.

A critical problem in an STM configuration is that an atomically sharp tip is required. The tip is normally made of tungsten, platinum, or platinum-iridium. Its shape dramatically affects the measurement results. In order to obtain atomic resolution, the tip radius is expected to be as small as an atom. However, it is impossible to obtain this using any of the tip sharping methods[209–211] that are currently used, e.g. cutting, grinding, electrochemical etching, and ion sputter thinning. The smallest tip radius which is sharpened by these methods is greater than tens of nanometers, and hence would appear unsuitable. Thanks to the principle of the exponential decrease of the tunnelling current with increasing distance from the surface, an atomically sharp tip could be obtained automatically. No matter how sharp the tip becomes when prepared by the above mentioned methods, there is a small protuberance of atoms in the front of the tip, the outermost atom dominates the tunnelling current due to the exponential dependence on distance. Other atoms, although just one or a few atomic sizes behind the outermost one, carry little of the tunnelling current. Therefore, the protuberance and the exponential decay of the tunnelling current as the tip-to-sample spacing is increased combine to provide a tip that is in effect less than 1 nm in size.

There are some other important problems associated with constructing a STM; for example, an accurate coarse driving unit is required to bring the tip from a millimetre distance to a few nanometers from the sample without colliding into the surface; again the tip-to-sample distance has to be stabilised on a sub-ångström scale by the construction of elements with dimensions in at least the centimetre range. This causes a mismatch over several orders of magnitude since the construction elements such as the tip, the scan unit, the sample holder, and the interconnecting base systems are affected by vibration and thermal drift. To solve these problems an ideal method is to reduce the number of construction elements to a minimum and to make the instrument as small, compact and rigid as possible. Many researchers[82,205–208,213] have achieved this objective.

A schematic diagram of a commercial STM system[212] is represented in Fig. 1.25. The system has three sections. The first section is the microscope which consists of the basic structures as described previously and a microme-ter-adjusted X-Y stage for sample positioning. Because of its compact and rigid construction, the microscope may be less than 250x150x150mm in size and a few kilograms in weight. The second section is a control unit which consists of (i) a X-Y board for providing a high voltage to the X-Y piezo-elec-tric transducers; (ii) a Z board for providing the high voltage to the Z piezo-electric transducer, a bias voltage to the sample and the tip, and for detecting the tunnelling current; and (iii) the high voltage power supplies. The third section of the STM system is the computer workstation which manipu-lates all the operations. Two monitors are provided, one is for topography image visualisation and the other for displaying the operating menu. The high speed special purpose digital signal processor carries out all the signal processing in relation to data filtering, image processing, topography charac-terisation and operation of the feedback control process. Additionally, some accessories such as a simple optical microscope and a video printer or a dot matrix printer are required in order to successfully observe the tip approach-ing the sample surface and for printing or hardcopying of the assessed 3-D surface topography.

1.5.2.2 Atomic Force Microscope

As is mentioned above, a tunnelling current is generated between two conductors. Thus it prevents the application of the STM to the measurement of non-conductive surfaces such as biological objects. To overcome the drawback of the STM, the AFM, which is a derivative of the STM was developed by Binnig et al. in 1986.[83] The AFM can be used for the measurement of either conductive surfaces or non-conductive surfaces. One of the earlier AFMs developed by Binnig et al.[83] is schematically represented in Fig. 1.26. A basic difference of the AFM from the STM is that a cantilever

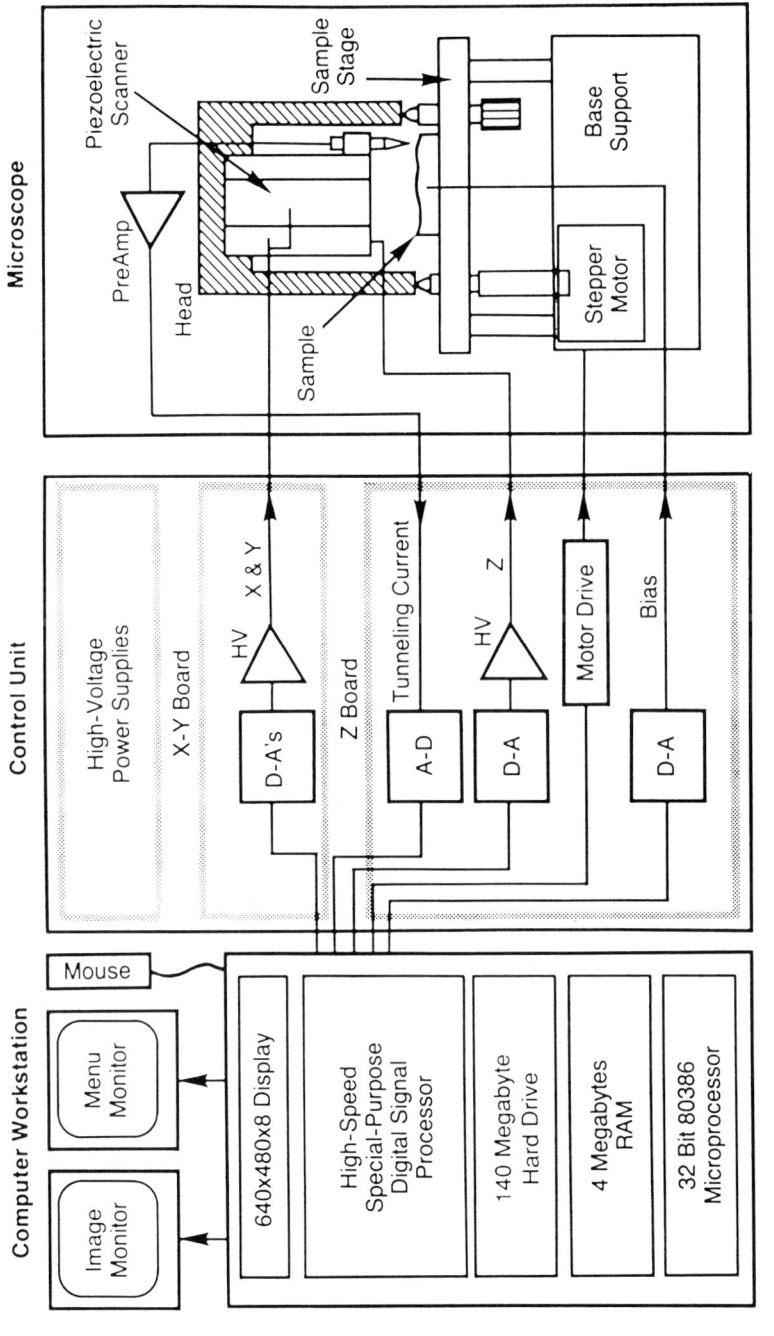

Figure 1.25 Schematic diagram of a STM system (from Digital Instruments)[212]

beam, to which an insulating or a conductive tip is attached, is used in between the sample and the STM. The tip makes contact with the sample surface, and an inter-atomic force exists between the tip and the surface. As the tip scans across the surface, the inter-atomic force changes due to the variation of the surface topography, and this force deflects the cantilever beam. By monitoring the change of the inter-atomic force or the deflection of the cantilever beam, the topography is obtained.

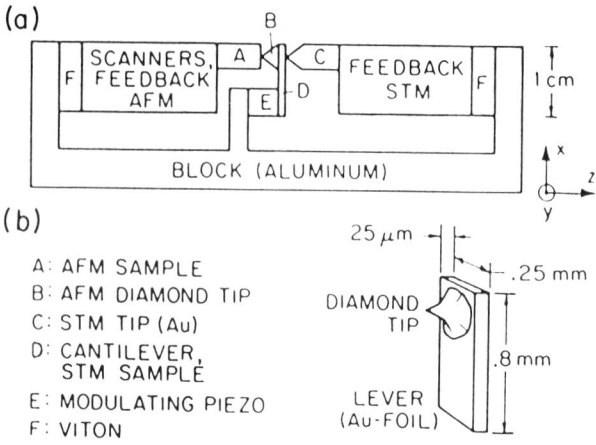

Figure 1.26 Schematic diagram of a AFM (from Binnig) [83]

Basically the AFM is similar to the stylus instrument, because both are contact measurement devices with a very sharp tip, both have cantilevers to sense signals and both need transducers (STM or LVDT) to pick up signals from the deflection of the cantilevers. For some newly developed AFMs, [200,203,208] a laser interferometer is used instead of the STM to sense the deflection of the cantilever. In principle, this is more similar to a stylus instrument – the Form Talysurf, for example, uses a laser interferometer instead of the LVDT. Besides the mechanical structural differences of the tips and the cantilevers (a more refined cantilever is required for the AFM), the major difference of the AFM from the stylus instrument is in the signal acquisition. The stylus instrument uses no feedback control mechanism, and the topography height is directly obtained from the output of the transducer (LVDT or a laser interferometer).

The AFM operates in four different modes which relate to the connections of the two feedback circuits shown in Fig. 1.26. The topography height is not directly obtained from the output generated by the deflection of the cantilever in the four modes; it is derived from the signal used to control the two feedback circuits.

- In the first mode the sample is modulated in the Z direction at a resonant frequency (several KHz) of the cantilever, the force F between the sample and the tip deflects the cantilever. This in turn modulates the tunnelling current which is used to control the AFM feedback circuit and maintain the force, F, at a constant level.

- In the second mode, instead of driving the sample, the cantilever is driven at its resonant frequency (>2KHz) in the Z direction with an amplitude of 0.1 to 10 ångström. The force F changes the resonant frequency of the cantilever and results in changes in both the amplitude and phase of the ac modulation of the tunnelling current. The change of amplitude of the ac modulation of the tunnelling current is used as a signal to drive the feedback circuits in this mode.

- The third mode is similar to the second mode, which uses the phase signal to drive the feedback circuits.

- In the fourth mode, the feedback circuit in the AFM side is controlled by the tunnelling current in the STM. The system maintains the tunnelling gap at a constant level by changing the force on the tip. Alternatively, the two feedback circuits can be reconnected in such a way that the AFM sample and the STM tip are driven in opposite directions with a factor α (ranging from 10 to 100) less in amplitude for the STM tip in the fourth mode.

It has been reported in the literature[208] that the results obtained by the fourth mode are superior to those obtained by the other three modes.

1.5.2.3 Properties of STM and AFM

As mentioned above, the vertical and lateral resolutions of the STM can be as high as 0.01 Å and 1 Å respectively. Atomic structure can be resolved to such high resolutions both vertically and laterally. As for the resolutions of the AFM, since it is a contact measurement, it is impossible to get the same effect as the STM in which the single outmost atom plays the most important role in the measurement. Thus the resolutions of the AFM in both directions are somewhat lower than those of the STM. About 1 Å resolution in the vertical direction and several nanometer resolutions in the lateral direction were claimed by some researchers.[83,94,203,213] The maximum size of a scan area of the STM/AFM is mainly dependent on the linear range of the X and Y piezo-electric transducers and the structure of a tripod or a tube scanner. At present, a size of about 100x100 μm is available for some commercial instruments.[212,213] The vertical range of the STM/AFM has been conspicuously absent in reports by authors in their earlier relevant papers.[81–83] In principle, if the constant current mode is adopted for the STM, the vertical range of the STM/AFM mainly relies on the linear range of the Z transducer and the structure of the scanner as well. A 5 μm vertical range of the STM and the AFM is now claimed.[212]

Owing to the very rigid structure of the scanner and the good frequency response characteristics of the piezo-electric transducers, the scanning speed of the STM/AFM is very high. A topography image frame with more than 256x256 data points can be obtained within a few seconds. This makes sub-real-time measurement and visualisation of surface topography possible.

The STM was originally developed for the study of surfaces under clean, well controlled conditions in a ultra high vacuum condition. Recently this has changed and it is understood[94,196,201,212] that the operation of the STM is no longer restricted to an ultra high vacuum. It also works at normal air pressure and would even work when immersed in a liquid. The reason for this is that the number of atoms left in a 10 Å^3 volume between the tip and sample is only 10^{-2} for ambient air pressure, so that it is still in a sense 'vacuum' tunnelling, and atomic resolution can still be achieved in air as well as in liquid environments. This property of the STM not only expands the application areas in biology, chemistry and materials, but also makes it possible to use a stand alone STM/AFM head to measure a large sample which does not easily fit into a standard STM sample stage.

In spite of the attractive properties of the STM/AFM, there are some shortcomings to the instrument. The preparation needed to adjust the probe to approach the sample within several ångströms is sometimes time consuming. Since the constant current mode assumes that whenever the feedback control system is adjusted to keep the constant current the surface height is determined, an error may be introduced when the probe is above a boundary of two atoms because the tunnelling effect is different for an atom and a boundary. On the other hand, if the topography height is obtained directly from the feedback voltage applied to the Z transducer, it does not guarantee that there is a linear relationship between the voltage and the displacement of the Z transducer in a wide measurement range. In addition, there are some limitations in relation to the geometric status of the probe.[214,215]

Since the invention of the STM and the AFM, some other types of scanning probe microscopes have been developed. The *scanning capacitance microscope* (SCM) is one of them. Although the SCM has not been commercialised, it has potential use in engineering because of its non-contact measurement capability and versatility.

1.5.2.4 Scanning Capacitance Microscopy

The use of capacitance techniques to measure surface topography was first proposed by Sherwood and Grookall[219] in 1968. They used a probe electrode to approach a conductive surface and to detect the capacitance between them, the capacitance being inversely proportional to the distance between them. Since the probe electrode they used was large, it could only detect the capacitance roughness height which was similar to the average height of the surface rather than true topography height. Brecker *et al.*[220] developed a similar capacitance sensor for workshop use in 1977. The probe electrode was cased in an insulator material which was in contact with the measured surface and used as the skid. It was more likely to be able to detect the waviness or envelope of the surface rather than the roughness topography. Some other types of capacitance devices for surface topography measurement were proposed by Liberman *et al.*[221] and Garbini *et al.*[222] These devices use skids and use, in effect, contact measurement techniques. Since all the above mentioned capacitance devices used large size probe electrodes in one or two horizontal dimensions, they were difficult to use to measure roughness topography in the corresponding dimension. Therefore, strictly speaking, they are not examples of capacitance microscopy in the scanning probe sense.

The first capacitance microscopy in the scanning probe sense was developed by Matey and Blanc.[216] A diagram of the capacitance probe is shown in Fig. 1.27; the electrode is attached to the edge of a diamond stylus which is mechanical contact with the samples. The stylus length is about 5 µm, the stylus width (into the page) is about 2.5 µm, and the thickness of the electrode is about 0.15 µm. The scanning system was built by modifying a commercially available VideoDisc player; it, therefore, carried out the scan in a spiral path, i.e. in polar co-ordinates, while the image display was shown on a CRT in the rectangular co-ordinates. Since the surface topography measurement mechanism with this scanning capacitance microscopy still used the traditional technique, the measurement range was limited. Again contact measurement may damage sample surfaces.

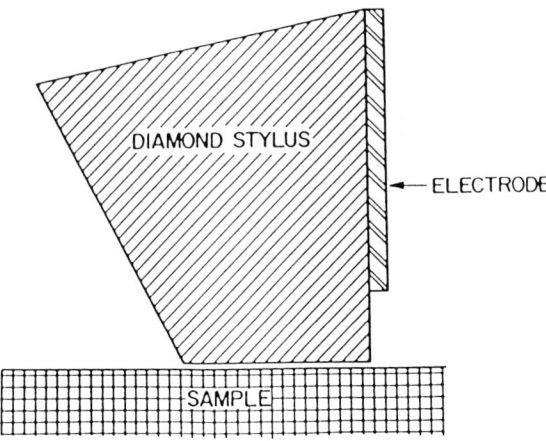

Figure 1.27 Schematic diagram of a capacitance probe (from Matey)[216]

A very promising scanning capacitance microscopy, which is the most similar to the STM in measurement principle, was developed by Bugg and King.[217,218] A capacitance probe which was a 20 μm diameter tungsten wire etched towards the tip to a diameter of about 1–2 μm was designed to have a non-contact scan over the sample surfaces as shown in Fig. 1.28(a). The surface height is not directly read from the capacitance between the tip and the sample. It is obtained from detecting the displacement of the capacitance probe driven by the servo control system (Fig. 1.28(b)). However, the increment of the capacitance between the tip and the sample induced by the change of the distance between them due to the movement of the X-Y stage rather than the actual capacitance is used to be the control signal fed to the control system. The measurement process is that the sample is put on a piezo stack which sits on the X-Y stage and is sinusoidally vibrated in the z direction. If the X-Y stage is static, the differential capacitance at the modulation frequency would be kept constant and a zero increment signal is fed into the servo control system to maintain the position of the wire tip with respect to the surface. As long as the X-Y stage has a movement, the capacitance changes, then the increment of the capacitance makes the servo control system drive the z transducer hence the probe. The movement of the probe is detected by the LVDT and forms the topography height in one point. As the x-y stage moves in a raster fashion, the probe scans over the surface and is kept constant distance from the surface through the capacitance detector and the servo control system. The surface topography is then obtained from the series reading of the LVDT. By adopting this technique, the scanning capacitance has a high vertical measurement range of up to 1mm, and a large measurement area of up to tens of millimeters. The vertical resolution of such a scanning capacitance microscopy is 5nm, depending on the separation distance between the sample and the probe, and the conductivity of the sample. The finest spatial resolution is down to 1–2 μm, which is determined by the probe tip diameter and the probe-surface separation. Both conducting and dielectric samples can be measured using this method.

These characteristics render SCM suitable for measuring engineering surfaces. However, there are some problems preventing it from being used in practice. One of the most important problems is the influence of mechanical noise. Also inhomogeneity in conductivity of samples may have influence on the measurement. Once these problems are sorted out, the SCM may become an important surface measurement instrument in engineering applications, especially in applications where non-contact measurement is required.

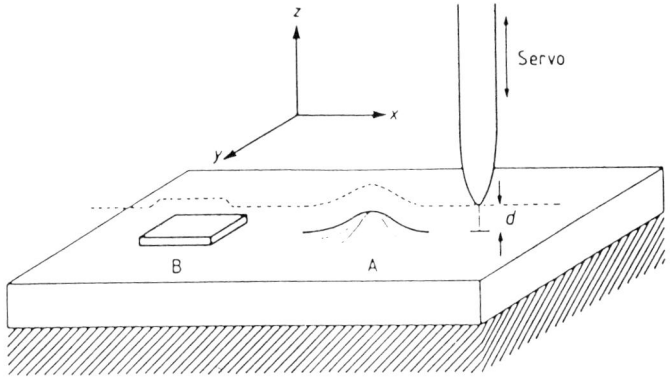

(a) Schematic diagram of the probe and measurement

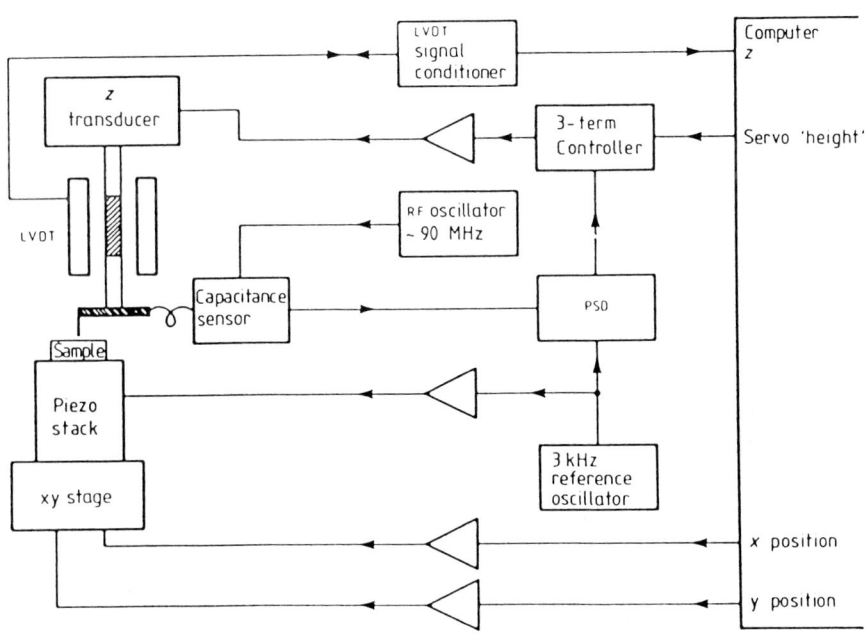

(b) Block diagram of the main components

Figure 1.28. Schematic diagram of a scanning capacitance microscope (from Bugg)[217]

1.6 GENERAL COMMENTS ON THE DIFFERENT TYPES OF INSTRUMENTS

In reviewing 3-D surface topography measuring techniques, it is not possible to say which technique is the 'best', because no one technique has all-embracing properties. Each has its own advantages and disadvantages in particular applications. In addition, users have different requirements for their specific applications and have their own instrument budgets. However, it is possible to state which is the best instrument to meet a specific application, and this, of course, depends on the user's requirement. In order to provide a guide to the strengths of different techniques, some of the important characteristics are here summarised, together with comments pertaining to the different types of instruments.

1.6.1 Measurement Range and Resolution

Measurement resolution and range as well as a derivative of the two – the ratio of the range to the resolution are very important characteristics of 3-D instruments. The first two characteristics of the instruments determine the capability to resolve the smallest object and its field of application e.g. biological or optical or mechanical engineering. The third characteristic indicates the capability to perform integrated measurements, that is, to measure more than one surface feature at the same time, for example, surface roughness and form (these effects are separation related). The larger the range/resolution ratio, the greater the possibility of integrating many features of the geometry into one measurement.

To compare the performances of the different types of instruments, an amplitude-wavelength plot of some instruments is drawn in Fig. 1.29. In the figure the two axes represent the resolution (towards the origin of the axes) and the range (far from the origin of the axes) of the instruments both in vertical and horizontal directions. Each block in the figure indicates the working area of an instrument. Drawing two orthogonal lines from any working point P in the area vertically and horizontally, the intersections P_b and P_l at the bottom and left hand ends give the resolutions at this working point and the intersections P_t and P_r at the top and right hand ends give the working ranges. The length of each line is an indication of the ratio of range to resolution in corresponding directions. The greater the length, the bigger the ratio.

It is seen that the STM/AFM measuring system has the highest resolution in both directions; however, the measurement range is small. This factor indicates that the STM/AFM is mainly suitable for measuring material and biological surfaces on the atomic or nanometer scale.

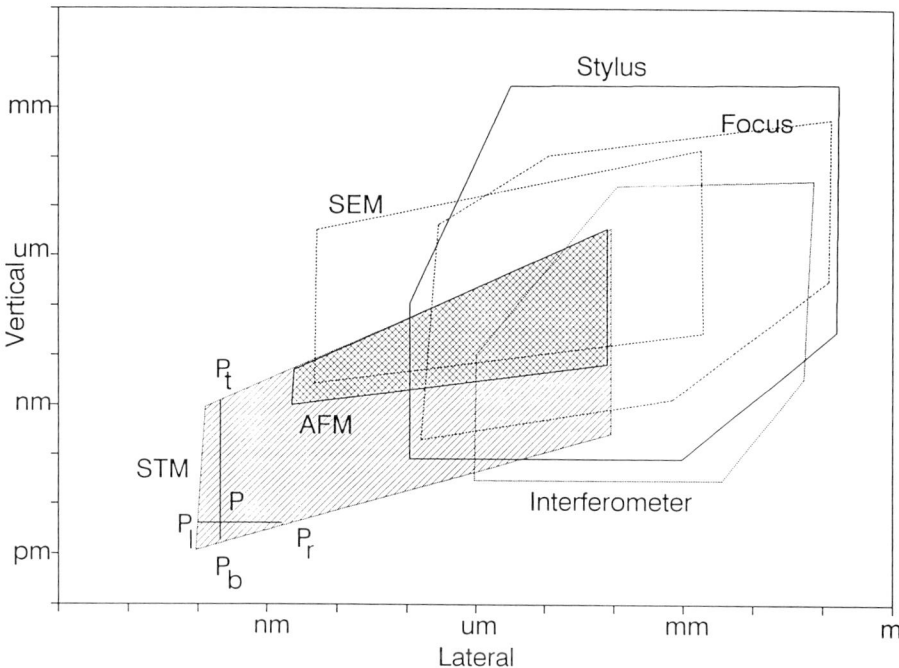

Figure 1.29 Estimated amplitude-wavelength plot of 3-D systems

The stylus instrument has a large vertical range, although it can have the highest vertical resolution to sub-nanometer, it is best suitable for measuring engineering surfaces at micron or sub-micron scale. The large ratio of range to resolution in the horizontal direction allows the stylus instrument to have an integrated information gathering process for roughness, waviness, form and other topography features left by manufacturing processes.

The focus detection instrument has a slightly inferior resolution to the stylus instrument in both the vertical and horizontal directions. However, its measurement range in the vertical direction has passed through the barrier of the wavelength of the incident light which is still a handicap effective for other optical instruments. The property of large vertical and horizontal range of the focus detection instrument makes it suitable for measuring surfaces similar to the stylus instrument.

The interferometer has the highest resolution (to sub-nanometer level) in the vertical direction, but the horizontal resolution is not comparable with this. It is even inferior to that of the stylus instrument. The vertical range is small due to limitations of the wavelength of the incident light. This makes it mainly suitable for measuring fine surfaces, such as optical and electronic circuits on the nanometer or the sub-micron scale.

The resolution characteristic of the quantitative SEM is the reverse of that of the interferometer. Its highest horizontal resolution is up to several nanometers, while the vertical resolution is never higher than this. The vertical range is also small. Therefore, the SEM is still useful for obtaining a topography image rather than quantitative 3-D information.

1.6.2 Measurement Speed

Considering measurement speed, although more than several tens of thousands of data points have to be sampled to form one frame of a surface topography image, the STM/AFM and the interferometer can collect one frame of an image within a few seconds due to their better frequency response characteristics. A sub-real-time measurement can be realised. On the other hand, the stylus instrument takes a long time to get a frame image because of the poor frequency response characteristics of the stylus-spring system. The frequency response characteristics of the X-Y stage is an important limiting factor in the data acquisition speed. Theoretically, the measurement time is proportional to n^2, where n is the number of points in one side of a square area mapped. That is to say, the time required to measure a square area does not increase linearly with the number of points n; it increases with n by a power of 2. Even if on-the-fly measurement is adopted, the stylus instrument could take at least 10 minutes or more to get a frame of an image with 256x256 data points. This is perhaps one of the most important reasons why the stylus 3-D measurement technique has not been widely accepted in industry. Although the scanning process of the focus detection instrument is the same as the stylus instrument, the former is faster than the latter because it is free from contact measurement and a mechanical dynamic system presented by a stylus-spring structure. However, its speed is still limited by the frequency response characteristics of the X-Y stage and the focus detection feedback control system; a few minutes is required to get a frame of an image with 256x256 data points.

1.6.3 Problems

The main problems associated with the stylus profiling technique are the contacting measurement, the influences of finite tip radius, stylus shape, stylus load and the dynamic characteristics of the stylus on measurement results, and the long time necessary for 3-D topography data acquisition. The focus detection technique is sensitive to surface inclination, reflectivity of samples, microgeometry and impurities of the samples. A disadvantage of the interferometry technique is that its vertical measurement range is very small – this limits the breadth of applications of such instruments. A uniform optical constant of the sample surface is also important for interferometry instruments. The SEM technique is more favourable for surface topography visualisation rather than for 3-D quantization, although quantitative techniques and commercial SEMs are available. The STM/AFM has its limitation in mapping large scale sizes and height due to its small vertical and horizontal measurement ranges.

1.6.4 Application Areas

In general, each technique or instrument has its own favourable application areas. The stylus instrument is best suited for measuring engineering surfaces. It is the most widely used instrument in manufacturing, production and metal machining, and in wear, friction and lubrication research. It is accurate, robust and even portable; therefore, it is a routinely used instrument in workshop practice and research. Moreover, it provides a standard technique to be referenced by other techniques due to its unambiguous definition through national and international standards.[148,149]

The instruments which are based on focus detection methods and feedback control mechanisms can provide wide measurement ranges and higher resolution, and are acceptable to the engineer. Except for some marginal conditions, a focus detection instrument has similar characteristics to a conventional stylus instrument, and can therefore be used in most applications where the stylus instrument is applicable. Furthermore, due to the advantage of non-contact measurement it is particularly useful where the stylus instrument could lead to damage of a sample surface and/or or the stylus tip itself. For example, it is best suited for measuring soft materials (soft metals, rubber, coal, paper, magnetic tape and liquid), surfaces with soft coatings (surface refined sheet metals, magnetic discs), very hard materials (hard metals, ceramics), elastically deformable workpieces (foils, films) and fine structures for which a conventional mechanical stylus cannot provide a 'true-to-form' representation. Compared with other methods of focus detection, the confocal method is the most suitable for constructing a scanning microscope, due to its ability to generate non-invasive serial optical sections of labelled specimens with a virtual absence of out-of-focus blur. This is of benefit to biologist and biomedical scientists, assisting them to understand the structure of individual organs, tissues and substructures hidden and the relationships and connections between these structures. Therefore, CLSM is now very popular in biology, biomedicine and chemistry.

The interferometric instrument is a high accuracy, high resolution, non-contact and low measurement range instrument. It is most suitable for use in precision engineering, especially in optical engineering, where the stylus and focus detection instruments are disadvantaged due to the contact measurement feature of the stylus method and the lower resolution in the case of the focus detection method. It has been proved that this kind of instrument has been successfully used in measuring diamond machined parts, transparent film surfaces, ball bearings, magnetic media and read/write heads, polypropylene surfaces, super-polished optics, lens moulds, fibre optics, and other optical and precision surfaces.

Unlike all conventional SEMs, currently commercially available SEMs are equipped with microcomputers and the sizes of the instruments are significantly reduced. The digital image processing function is provided by most current SEMs. Quantitative 3-D SEMs are available.[190–192] Conventional analysis for surface topography, such as statistical and spectral analysis, can be carried out using this instrument. Furthermore it shares with the confocal instrument the ability to analyse segmentation and oblique sections of surfaces and to reconstruct the 3-D image from a series of image sequences. The SEM is a traditional and routinely used instrument in material and metallurgical sciences, and is also widely used in biological science. It is reliable and efficient for use in observing surface structures of materials, to measure size, shape densitometric parameters of grain and boundary structures and to detect surface fracture.

The applications of STM/AFM have developed far beyond original expectations. The fact that it has higher resolution than any other instruments in both vertical and spatial directions is not the only advantage it possesses. The fact that it can be operated at ambient air pressure and even in liquid environments is of equal importance. This instrument is not merely used to observe atoms and fine details of molecules, it may be used as a structure modifying tool in the nanometer and subnanometer scale to manipulate single atoms, e.g. to 'pick up' an atom from a surface and to move it to other place of the surface by changing the tunnelling conditions. Compared with other 3-D instruments, the STM/AFM is superior for subnanometer measurement and widely accepted by science and engineering. In biological science the STM/AFM is a useful tool for visualising DNA molecules and viruses. Biochemical operations are now possible due to the ability of the tip to deposit a small amount of material on a surface. In chemical engineering the STM/AFM allows investigation of structures formed by adsorbates and chemisorbates or chemical surface reactions. In material science crystal structure, growth and crystalline formation can be observed by the STM/AFM. In mechanical engineering the STM/AFM can be used as a micropositioner which enables the tip to be positioned in a specific location on a surface with atomic accuracy. It is also a micro-scale measurement instrument. Lithography is obtainable by deliberately etching a surface with the STM/AFM which leads to, in another application, development and manufacturing of microcircuits. There are probably many applications of the STM/AFM which are still under investigation.

1.7 CONCLUSIONS

In this Part the state-of-the-art of measurement instruments and techniques of 3-D surface topography have been reviewed. The proposed techniques, based on different optical and scanning probe principles for measurement of 3-D surface topography, have been reviewed drawing on wide ranging literature. As a consequence, it is difficult to include all of them. However, those introduced are representative only of the papers published and are intended to point out the most promising devices and commercially available instruments. The appendix lists the technical specifications of a number of commercially available 3-D instruments with the addresses of the corresponding companies.

It is clear from the review that

(1) there are many commercial instruments available for the quantitative measurement of 3-D surface topography.

(2) these instruments are based upon different measurement principles, which have been discussed.

(3) the working range of the instruments covers from atomic dimensions to large machine component size.

(4) the instruments are complementary in measurement, each of them having its own ideal application area.

(5) sub-real time measurement is possible using some optical and scanning probe microscope instruments.

(6) 3-D measurement is not rare, the instruments having already been widely applied in many disciplines such as mechanical, manufacturing, electronic, optical, chemical, biological, biomedical engineering/sciences.

3-D measurement has developed over the last two decades, but in recent years progress has been rapid due to the introduction of powerful microcomputers and other advanced measurement and analysis techniques. It is expected that the future development of the techniques will be in two directions. The first, probably the more important one, is the application to a wide range of products with more comprehensive interpretation of measurement results in engineering and science. The second is the development of the instruments themselves. It is believed that the future development of instruments will be mainly concentrated on enlarging the ratio of range to resolution and increasing the ability to minimise the influence of ambient environment so as to achieve in-process 3-D measurement rather than simply increasing resolution. Non-contact techniques will be further investigated, in future, in preference to contact techniques.

Part II

THREE-DIMENSIONAL SURFACE TOPOGRAPHY – REVIEW OF PRESENT AND FUTURE TRENDS

W P Dong, E Mainsah,
K J Stout and P J Sullivan

THREE-DIMENSIONAL SURFACE TOPOGRAPHY – REVIEW OF PRESENT AND FUTURE TRENDS

In this Part, we present the results of a recent survey conducted as part of an EC project aimed at developing an integrated approach to 3-D surface topography assessment. The survey was carried out among surface topography researchers, manufacturers and users in both academia and industry in seven European countries (the UK, France, Germany, Sweden, Belgium, the Netherlands, and Denmark). The academic institutions include universities, polytechnics and research institutes. There was a broad spectrum of industries covered in the survey – instrument manufacturers, car manufacturers, chemical and steel plants and machine tool manufacturers – further evidence that surface topography analysis is multi-disciplinary and now permeates numerous areas of interest in applied science.

A statistical analysis of the questionnaire replies – which were evenly distributed between industry and academia – shows quite clearly that optical techniques are now more widely used and accepted in industry than previous researchers have implied. A further analysis of the results in terms of user expectations and the perceived relevance of functional parameters in key application industries is a good indication of the present state of 3-D surface topography – from a customer view point – and we believe that this is also a pointer as to how one can expect 3-D surface topography to develop in the future.

2.1 INTRODUCTION

Over the past decade, surface topography analysis has moved from a purely two dimensional (2-D) approach, to a 2-D dominated 3-D and 2-D hybrid – one where 2-D analysis is still significant but where three dimensional (3-D) analysis is gaining increasing acceptance and is widely seen by many researchers as the way forward.[1] The 3-D approach is a logical conclusion based on many factors; it is acknowledged that any functional assessment of an engineering component must take into account the interaction[2] of that component with others – must take into account the three dimensional nature of all components. Computer power, a crucial factor due to the large volume of data processing involved, is improving all the time and prospects for carrying out 3-D data analysis in a real-time environment are getting better.

Several authoritative surface topography instrumentation surveys and reviews[3–7] have been conducted and have largely concluded that stylus data acquisition methods are still the most widely used, despite the established drawbacks of the stylus method (stylus geometry effects, and deformation caused to relatively soft surfaces) and the widely-acknowledged possible improvements obtained with the use of optical methods. Optical techniques have been slow taking off because of their industrial limitations – cost, surface amplitude range, frequent need for surface preparation, the need for close environmental control, and the fact that a skilled operator is very often required.

2.2 AIMS OF THE SURVEY

The lack of a co-ordinated approach to the development of 2-D surface topography analysis has resulted in numerous practical, communication, economic and developmental problems in 2-D surface topography instrumentation, characterisation and interpretation, all of which have been well documented.[8,9] The EC project[7] of which this questionnaire was a part, was conceived to try and develop a consolidated approach to 3-D surface topography analysis early in its development, so as to avoid the mistakes inherent in the earlier 2-D development. This is even more important as the EC gets closer and closer, and hence a logical step in the establishment of ground work for standardisation was to conduct a survey of present trends so that this could then be used as a basis for further work.

It was intended to establish a number of points:

- To investigate the impact or expected impact of 3-D surface topography on shop floor data acquisition.
- To determine the breadth of surface topography application industries.
- To investigate the range of instruments available with respect to their data acquisition principles.
- To investigate the actuation mechanisms employed in surface topography instruments.
- To investigate the data sampling and digitisation procedures used in surface topography.
- To investigate the range of 3-D parameters used.
- To obtain users' perception of the functional information which can be obtained from the parameters in use.
- To obtain information on the variety of characterisation approaches currently adopted by researchers in academia and industry.
- To ascertain the functional information which users would like to get from 3-D parameters.

2.3 MAIN FINDINGS OF THE SURVEY

2.3.1 3-D Versus 2-D Analysis

Some people have long argued, perhaps reasonably, that 3-D analysis needs to remain a laboratory tool for much longer, before industry can reasonably be expected to use it on the shop floor. It has been argued that 2-D analysis, simpler in comparison, is not yet fully understood and that the parameters generated have little functional significance,[9] and the problems of 2-D analysis will need to be solved before sensible 3-D analysis can be carried out. Others have argued, however, that the use of 2-D analysis to predict the functional performance of a machined part is itself fundamentally flawed,[1,8] since interaction with other machined parts is itself very much a 3-D phenomenon. Any meaningful analysis must, therefore, take this into consideration.

Three dimensional analysis has also introduced a very important functional characterisation tool: visualisation. Although effective characterisation must include an objective analytical aspect, there is no doubt that the visual representation of a surface is a vital feature from which the engineer can begin to make important decisions about the probable function of a specimen. Through visual inspection, the engineer can observe surface features (craters, lay, scratches, summits) in greater detail by zooming in on them and through the use of the visualisation plots – contour plotting, inversion, truncation, axonometric plotting, intensity plotting – can make well-informed predictions about specimen function. This facility is conspicuously absent in 2-D analysis.

The results of the survey would seem to indicate very strongly that the proponents of 3-D analysis are winning the argument. A good 89% of respondents to the questionnaire considered that 3-D analysis was important for their application, with less than 10% registering a negative answer. It is significant to note that those who felt that 3-D analysis was not important in their work all cited the slow speed of data acquisition (which was viewed as impractical for normal shop floor operations) as the causative factor in their negative response.

2.3.2 The Scope of Surface Topography

3-D surface topography research is multi-disciplinary, including in its scope instrumentation, measurement, data processing, manufacturing, tribology, wear etc, and hence it is vital that there is maximum co-operation between scientists and engineers from different fields in order that a comprehensive and widely-acceptable 3-D surface roughness standard can be developed.

The application industries that were represented in the survey included instrument manufacturers, car manufacturers, chemical and steel plants, and machine tool manufacturers – this is further evidence of the broad size of the potential market for surface topography.

2.3.3 Instrumentation

Currently, three main types of instruments are commonly used –

- stylus-based profilometric data acquisition systems,
- optical-based data acquisition systems,
- the scanning microscope in its various derivatives.

2.3.3.1 Stylus-based Systems

Stylus-based instruments have the longest history of use in surface data capture and for many years have been the most widely used in industry, especially in automotive and metal-related industries. 3-D stylus instruments have been directly developed from 2-D instruments by the addition of an extra translational degree (y), perpendicular to the (x, z) plane. This is realised in two ways.

(i)   One translational table is added in the y direction. (63.5% of systems in the survey effected motion using this application.)

(ii) Two perpendicularly mounted translational tables are used whilst keeping the stylus stationary.

Presently, almost all stylus instruments are based around the microcomputer, that is, the measurement system comprises two parts – the first is the traditional stylus instrument responsible for the measurement, amplification and output of the signal; the second part is the microcomputer responsible for motion control of the measurement procedures, processing of data, characterisation of the surface and display of the necessary results including parameters and graphics. These two parts are connected either by a parallel or serial interface, and by an analogue-to-digital (A/D) board which is used by the microcomputer for digitising data from the analogue signal.

2.3.3.2 Optical Systems

The rate at which optical systems are being used in 3-D applications is constantly increasing. The rate of increase is so high that the new role that optical systems are playing in 3-D topography can easily be under-estimated. A very significant result in this survey was the clarity of the increasing role of optical instruments. Conventional wisdom, based on earlier reviews, holds that optical systems (certainly in routine shop floor measurements) have a long way to develop. The survey found, however, that the role of optical systems might have been under-estimated. The results show that 50% of the respondents use an optical system, with only 45% using stylus-based systems, while 5% opted for the scanning microscope or its derivatives.

Optical 3-D surface topography measurement systems are increasingly being accepted in academia and industry because of the great advantages associated with non-contact measurement. Due to the non-contact nature of the measurement, no physical contact is made with the specimen, thus avoiding damage to the surface. This makes optical systems ideal for use in a wide range of applications so that in areas such as optics, integrated electronic circuits and painting, optically-based instruments are more widely used than stylus ones. Another significant feature of optical systems is that they have a higher vertical resolution than stylus-based ones but the main draw-back is that the measurement range of the former is smaller than that of the latter.

Optical systems are based on different optical principles. The most popular are based on the following three principles:

- interferometry (26% of respondents),
- light scattering (15% of respondents),
- focus detect (30% of respondents).

In *interferometry*, surface asperities are measured by measuring the phase shift between a measuring beam reflected from the surface, and a reference beam. As the measuring beam scans the surface, the changes in surface height yield phase shifts in the reflected beam that are compared with the phase of the reference beam. The phase change of the reflected measuring beam is directly proportional to the path length travelled to and from the reflecting surface and hence the surface height itself. Interferometric systems have higher vertical resolutions, but there is a maximum roughness limit that can be measured, usually about half the wavelength of the incident light.

Since the optical phase shift of the reflected beam depends on the surface material as well as the height, an ambiguous measurement of surface profile can occur during the measurement of non-homogeneous surfaces. Examples of interferometric systems include the *Wyko Topo* and the *Zygo Zerodur*.

Light scattering methods utilise the scattering property of rough surfaces for the determination of a characteristic value for their roughness. The surface to be analysed is illuminated by an intensive beam of infra-red rays. A fraction of the radiation is scattered back, the angle of distribution of the scattered rays depending on the surface structure. With the aid of an optical system and a photo diode array, a cross section is made through this scattered beam and the intensity distribution is measured as a function of the angle of dispersion.

Light scattering instruments have lower vertical resolutions than interferometric ones, and normally only produce relative parameters rather than the absolute height of surface roughness; thus it is usually used for relative roughness measurements. They are also relatively cheap.

Focus detect systems are based on detecting the focus position of the incident light and hence the profile height. Typically, an optical focus stylus of 1 μm diameter replaces the tip used in mechanical instruments. A focus detector analyses the light reflected from the object. Depending on whether the focus is above or below the surface, the detector generates an appropriate signal for re-focusing the lens. The lens movement accurately follows the surface contour and is measured with a high precision inductive displacement transducer.

Focus detect systems have vertical resolutions which lie between interferometric and light scattering systems, but the vertical range of measurement is as high as 1 mm. Since the realisation of movement is similar to that of stylus systems, the horizontal resolution and range of the two systems are comparable. The accuracy of the system is influenced by local profile inclination, reflection behaviour, micro-geometry and surface impurities, all of which limit the application range of the system. Examples of focus detect instruments are *Perthen Focodyn* and the *Rodenstock RM600*.

2.3.3.3 The Scanning Microscope

Scanning microscopes including the SEM (Scanning Electron Microscope) and the STM (Scanning Tunnelling Microscope) are the highest precision surface topography measurement instruments currently available. They make possible the direct observation of atomic scale structures on surfaces, but their measurement range is very small. The main draw-back of the SEM or the STM is that specimens must be electrically conductive. Non-metal specimens can be examined by coating them with a thin layer of metal by a vacuum evaporation process. Examination of surfaces is not a quick process, since the preparation of the specimen takes some time. The cost of these instruments is also very high, and they are currently mainly used for research purposes.

Besides the above-mentioned measurement techniques, some other techniques such as capacitance,[10,11] pneumatic,[12] and ultrasound[13] have been recently developed. However, a few problems need to be overcome before these techniques can become more widely accepted and used.

The main features of these 3-D surface topography measurement systems are listed in Table 2.1. An important factor that makes these systems impractical for on-line measurement (as is the case for 2-D) is that specimens very often have to be prepared, and this adds time overheads to the measurement process. Operator skill is also a limiting factor. It is clear, therefore, that the development of an efficient and fast 3-D surface topography system that can be used for on-line measurement and monitoring of the production process is still a matter of considerable importance.

2.3.4 Digitisation – Range and Resolution

Since the computer has been incorporated into the traditional measurement system, that is, a digital technique has been adopted, the vertical resolution not only depends on the analogue output but also on the resolution of the A/D converter. Fig. 1 shows an example of the resolution relation between the vertical magnification (Vmag), the full range deflection (FRD) of the instrument and the bits of A/D converter.[14] It can be seen that

- at a fixed vertical magnification, the resolution increases twofold as the number of A/D bits increases by 1;
- vertical magnification determines the FRD, which in turn influences the resolution of the whole system.

Although the vertical resolution of such an instrument could be as high as 1 nm, the authors found that a resolution of about 0.01 µm is common for most stylus systems – perhaps due to the fact that larger magnifications or more A/D bits tend to result in a significant noise factor and also increase the cost of the system.

According to the survey, 48 % of the systems use 12 A/D converter bits, 17 % use 14 bits, 17 % use 10 bits, 10 % use 9 bits, and 9 % use 13 bits. At present, a 12 bit A/D converter probably gives the best value for money as it would appear to provide a balanced trade-off between cost and accuracy.

The horizontal resolution of the system depends on both the radius of the stylus and the accuracy of movement of the translational table or the gear box. For most systems this value was found to be about 1 µm.

The vertical range of such system depends on the range of the sensor located inside the pick-up of the instrument. This normally varies from 0.5 mm up to ±2 mm depending on the particular instrument. The horizontal range is limited by the range of the translation stage or the traverse length of the stylus. A range of about (100x100) mm is very common.

2.3.5 Levelling of Stages

Levelling of the specimen surface is important if a true representation of the surface is to be obtained. This factor is important in all types of instruments (although the effect is greatly reduced if the horizontal range is small). If the surface is not levelled, the measurement results may be greatly distorted as there is a reduction of the vertical and lateral measuring range and the reliability of the data becomes questionable. Surface levelling is usually implemented by manually levelling the stage, but software levelling procedures are also possible, for example, through software compensation for small inclinations of the surface.

Table 2.1: Features of 3-D surface topography measurement systems

	Stylus	*Optical*	*Scanning Microscope*
Vertical Resolution	High	High	Very High
Vertical Range	Large	Small	Very Small
Horizontal Resolution	Medium	High	Very High
Horizontal Range	Large	Medium	Very Small
Measurement Mode	Contact	Non-contact	Non-contact
Time for Preparing Specimen and Instrument	Short	Fairly Long	Long
Time for Measurement	Long	Short	Short
Cost of Instrument	Medium	Medium	High
Influence by Inclination of surfaces	No	Yes	Yes
Influence by Environment	Not so sensitive	Sensitive	Very Sensitive
Influence by Reflectivity of Surfaces	No	Yes	No
Dependence on Conductivity of Surface	No	Yes	Yes
Damage Surfaces	Easy	No	No
Suitable application areas	Car, Machine Manufacturing	Optical, Elec. and Painting	Material

Currently there are two main kinds of levelling stages; one type is motor driven[15] and the other is manually adjusted. Although 57% of the systems surveyed had adopted a levelling technique, flexible and low cost automatic levelling systems still need to be widely considered.

2.3.6 Specimen Relocation

In wear and tribological research, it is usually necessary that a fixed position on a surface be measured several times during its normal operation so as to observe tribological effects during the running life of the specimen. The most commonly-used and convenient way to relocate a specimen as evident from the survey, is by a physical fixture (53.5% of respondents). Some other auxiliary techniques such as optical, (28.5%) sensor, (7%) and camera (11%)

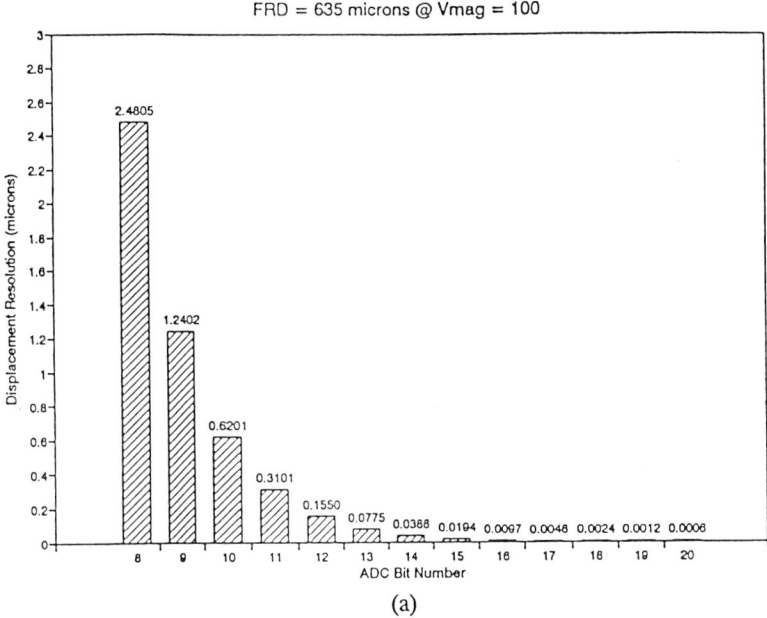

(a)

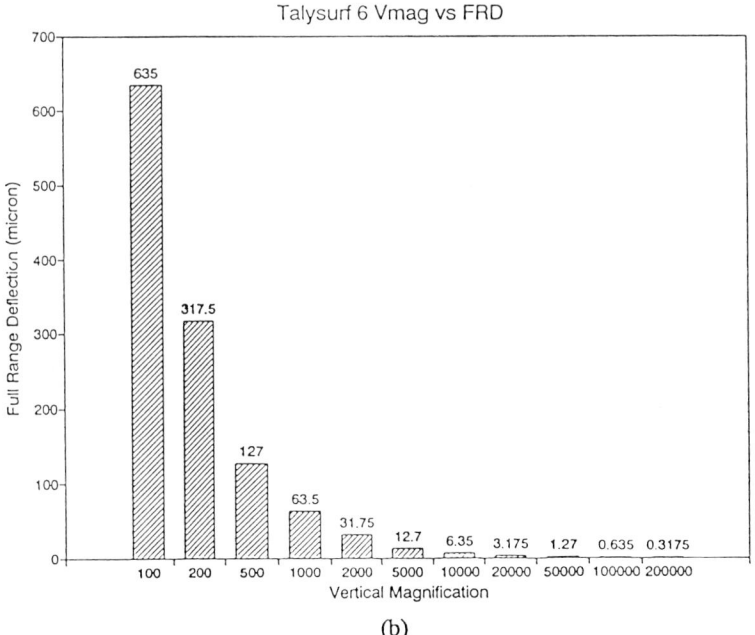

(b)

Fig. 2.1: Example of the resolution relation of stylus instrument (a) relation between vertical resolution and ADC number of bits, (b) relation between FRD and Vmag

are used to enhance relocation. A recent report[2] has shown that the use of a video camera can aid relocation and get repeatability to within 5 µm or less in both the X and Y directions.

2.3.7 Measurement Datum Plane Definition

In 2-D profile measurement, especially in stylus systems, the datum is vital to the measurement result since the output signal is related to the datum adopted. In most cases where a physical datum such as a skid datum or independent datum is used, the signal is, to some extent, distorted because the datum would not be an exact straight line. Since a profile is an entity in 2-D measurement, the error from the datum is just confined to this profile and the choice of cut-off (i.e. high pass filter) has little effect on the datum.

In 3-D surface topography measurement, the measurement datum is a plane. Since maintaining a physical datum in a plane is more difficult than maintaining it in a straight line, many 3-D systems do not use physical datums at all, that is, a skid (for stylus instruments) is not used. In this case, the movement plane of the X and Y tables (or gear box and Y table) becomes the measurement datum. Unlike 2-D profile measurement, high-pass filters are not necessary if the surface is measured trace by trace, otherwise the measurement datum may be composed of discrete profile datums, hence resulting in loss of some 3-D information.

2.3.8 Static and Dynamic Measurement

There are two main measurement modes in surface digital measurement systems. One is static, that is, the horizontal relative movement of the stylus (or optical spot) with respect to the specimen, and data sampling, are not simultaneous. After the relative movement stops, one data point is read by the analogue-to-digital converter (ADC) before the probe is moved to the next target point and the procedure repeated. This mode reduces the dynamic influences of measurement systems such as the frequency response and the dynamic characteristics of the stylus, but it is expensive in terms of measurement times. Also, backlash and elastic deformation of the driving device may affect the positioning accuracy due to the intermittent movement.

The second measurement mode is dynamic (or *on-the-fly*), that is, relative movement and data sampling are simultaneous. Whenever the displacement reaches one sampling interval, one data point is read *while* motion continues. The sampling procedure is normally performed by one of two ways.

- triggering by a timer inside the computer (when the velocity profile of the moving part is known),

- triggering by the output of a position sensor such as an optical or magnetic grating.

The latter mode is capable of higher positioning accuracy, and greater efficiency but the measurement speed is limited by the dynamic characteristics of measurement systems, especially the stylus. Since the stylus is actually a vibration system, a rough surface could act as an excitation source to induce vibration in the stylus, and fast traverse speeds could result in the tip of the stylus losing contact with the surface. Even in optical systems, the measurement speed is still limited by the dynamic response of the systems. For example, a focus detect system takes some time to adjust the focus position. The effect of the dynamic characteristics of the stylus on measurement still needs to be fully researched in order to improve the efficiency of dynamic measurements.

It was found that the data acquisition mode was fairly balanced between static and on-the-fly. In the sample, 56% of the systems used *on-the-fly* data collection and 44% opted for static data logging.

2.3.9 Data Logging Conditions

Generally, two of the following three parameters are variables in the sampling strategy – the sampling interval, the size of the sampling matrix, and the sampling area. The third parameter can be calculated from the other two.

The survey revealed that the sampling interval is the most widely used independent parameter followed by either the size of the sampling matrix or the sampling area. Since no standard for data sampling is available, the sampling interval and sampling area (or size of sampling matrix) adopted by respondents to the questionnaire cover a wide range of values (sampling intervals from 1 nm to 50 μm, sampling areas from (1x1) μm^2 to (20x20) μm^2, size of sampling matrix from 100x100 to 1027x128, with no clear indication as to the selection criteria for these parameters. The values, however, seemed to be application-dependent.

2.3.10 Digital Filtering

Filtering can be a great asset in visualisation, which in itself is a valid form of functional characterisation. There are various reasons why 3-D filtering is necessary:

(1) In order to separate areal roughness from areal waviness and form.

(2) In order to isolate and extract features of interest.

(3) In order to remove unwanted frequency components and noise disturbances.

According to the survey, 79% of respondents have adopted digital filtering techniques, so illustrating the importance of digital filtering. The techniques used include the use of convolutions, discrete Fourier analysis, parabolic curve fits, and curve fitting via averaged profiles.[16–20]

None of the above filtering techniques is a panacea for all filtering problems and, therefore, digital filtering techniques used in the analysis of surface topography need to be further developed so that standard practices can evolve on which a future 3-D standard could be based.

2.3.11 Characterisation Reference Datum Plane

The reference plane is important in assessing surface properties. It provides the measurement datum for all profiles in a 3-D array and from which surface parameters are calculated.

As in 2-D analysis, many kinds of reference planes are possible. The effectiveness and advantages of these reference planes have not been conclusively investigated. According to the survey, the least squares plane is very widely used (73% used the least squares plane, 12% used the arithmetic mean plane, and 15% used other methods of datum definition). The popularity of the least squares plane is probably due to the fact that the least squares plane is unique and is convenient to calculate using a digital computer. It is also an effective method of levelling areal data.

2.3.12 Characterisation and Parameters

A key problem in surface-related research is finding parameters that characterise surface properties in such a way that they correlate with its formation mechanisms and behaviour in a fundamental way.[7] A complete description that would be useful for predicting surface behaviour and for understanding the mechanism of its formation should include its chemical composition, micro-structure, residual stresses and topography. However, from an *engineering metrology* point of view, characterisation is not merely about description and measurement of surface topography; it is also about the interaction between topography and engineering requirements such as wear, lubrication, tribology, seals and thermal contact etc. It is well known[21] that some characterisation techniques such as statistical analysis, spectral analysis and time series analysis have long been used in 2-D profile characterisation.

Many researchers have extended these techniques to 3-D topography analysis directly, and many descriptors borrowed from 2-D analysis have been used in 3-D analysis. The characterisation methods cited in the questionnaire were statistical (35%), spectral (15%), functional (27%), time series (8%), fractal (8%) and Motif (7%).

Characterisation techniques can be split into two major groups, one comprising techniques that are scale-dependent (that is, the characterisation results depend on the measurement scale used), while the other group is made up of scale-independent methods (that is, the adoption of a measurement scale does not affect the characterisation results). Some of the methods in the first group are directly extended from 2-D profile analysis in which both theoretical models and parameters are adopted and extended to include 3-D analysis. Some, however, are specially proposed for 3-D topography and functional analysis.

2.3.12.1 Statistical Characterisation

Statistical characterisation[22–25] is divided into two parts; one is involved with the construction of statistical mathematical models (such as autocorrelation functions) to describe the random process and the other is concerned with the definition of statistical variables (such R_a, and R_q) to provide a quantitative description of the surface. These statistical parameters are useful for classifying surfaces according to the process of manufacture; for example, it is easy to identify turned and ground surfaces by the magnitude of these parameters. However, since the topography of most surfaces is very complex, the height distribution of the surfaces does not usually match the ideal random model such as the Gaussian distribution, and consequently, these parameters are subject to variation. It has been demonstrated that where pits and troughs exit, the variation of some of these parameters can be intolerable.[26]

2.3.12.2 Characterisation Via Spectral Analysis

Spectral analysis is a powerful tool for analysing surface topography that is dominated by periodic components.[4,17,18] A theoretical model, the power spectral density (PSD) function, can be built into the analysis, with the magnitude at each frequency representing the strength of this frequency component in the surface. The periodicity of significant components is easily identified from the PSD function as well. This technique is best suitable for representing the relative contribution of different frequency components, for example, for observing the magnitude relation between roughness and waviness.

2.3.12.3 Time Series Analysis

This technique is another useful method for analysing random signals.[27,28] Three kinds of time series models, AR (Auto Regression), MA (Moving Average), and ARMA (Auto Regression and Moving Average) can be constructed. The random topography data can then be fitted on to the model.

The coefficients of the chosen model may represent features and functional characteristics of the surface. The problems associated with this technique are

- it is difficult to decide the order of the models;
- it is only practically applicable to random surfaces;
- the interpretation of the coefficients is not straightforward.

2.3.12.4 Functional Characterisation

Functional characterisation directly relates surface topography to engineering applications,[29–32] so it may be more beneficial and more realistic in its approach to analysis than other characterisation techniques. The problem with functional characterisation is that the engineering requirements for a particular surface are still not fully appreciated and documented and there exist a knowledge gap between features which must be specified by the designer and those which can be generated by appropriate manufacturing processes. The lack of understanding results from the engineer's lack of information about the aspects of surface roughness which influence functional performance. Although some functional characterisation approaches such as the plateau surface model,[29,30] the truncation model,[31] and surface lay characterisation[32] have been proposed, not many functional parameters are included in national or international standards. Therefore, the development of 3-D parameters which are holistic in their description or representation of the surface topography as well as functionally significant remains an urgent need in industry, and a pressing challenge for those involved in surface topography research.

One of the major reasons for the upsurge in interest in the development of 3-D surface topography analysis techniques is that there is increasing realisation that 2-D profile analysis techniques cannot fully meet the needs of most engineering applications.

This is due mainly to some or all of the following:

(a) all engineering parts are surface-related rather than profile-related;

(b) the engineering applications of parts is concerned with functional properties and this has to be related to the characteristics of the 3-D surface topography;

(c) parameters calculated from a profile can not represent the characteristics of the surface because the sample size would need to be unrealistically large for profile characteristics to completely represent the areal characteristics.

As regards functional characterisation techniques, the following methods have been adopted and used in various applications:

- Area bearing ratio,
- Motif combination,
- Comparison with results of functional tests,
- Truncation,
- Visual characterisation,
- Morphological and
- Spectral.

Functional characterisation was seen as the most important aspect from an industrial view point, and there was a general need for a wide range of functional information pertaining to a wide group of engineering applications. Some of these are summarised below:

(a) Surface function, contact state of surface, bearing area.

(b) Oil volume, lubrication property of surface.

(c) Wear property.

(d) Frictional property.

(e) Anisotropic parameters, lay directional parameters.

(f) Plastic deformation, information for FE (Finite Element) modelling.

(g) Properties influencing optical usage.

(h) Height and portion of peaks and valleys of surface, ability to identify pits and troughs.

(i) The role of roughness and waviness.

(j) Ranges of parameters by which relative motion parts contacted are best matched and have the longest life.

(k) Ability to predict contact pressure in load.

(l) Harmonic content of the 1st and 2nd frequency.

(m) Imaging quality of optical surface.

(n) Surface deformation after laser welding and cladding.

(o) Characterisation of deterministic texture.

(p) Crater and rim parameters as formed by laser or electronic beam texturing.

(q) 2-D Power density spectrum and 2-D auto-correlation function.

(r) Separation of 3-D roughness from 3-D waviness and waviness from form deviation.

Since different engineering applications have different requirements for the functional properties of the surface, (Table 2.2)[7,38] an uncontrolled rush for functional parameters would result in little more than sheer confusion. One sensible approach which would limit the catalogue of unjustifiable functional parameters would be to classify functional requirements into several categories, and then include the classification and the criteria for the classification into a future 3-D standard together with a clutch of useful parameters.

Table 2.2: Functional performances of surface related to engineering applications

	Roughness	Waviness	Form	Lay	Laps & Tears	Chemistry	Metallurgy	Stress & Hardness
Wear	X	X		X	X	X	X	X
Friction	X	X		X	X		X	X
Lubrication	X	X	X	X	X		X	
Sealing	X	X	X	X				
Fatigue	X					X	X	X
Corrosion	X					X	X	
Thermal	X	X	X		X	X	X	
Electrical Resistance	X					X	X	
Magnetism						X	X	X
Reflectivity	X	X		X	X	X		
Cleanliness	X				X	X		
Coating	X	X				X	X	
Painting	X	X						
Plastic Deformation	X	X					X	X
Elastic Deformation		X	X				X	X
Assembly	X				X	X		X
Accuracy	X	X	X		X			X
Part Life	X	X			X		X	X
Joints	X	X	X	X				X
Creep	X	X		X		X	X	X
Load Carring	X	X						X
Fluid Flow	X	X		X				

2.3.12.5 Visual Inspection

Clearly, one cannot fully appreciate a surface by simply viewing a profile but by viewing the surface topography. Although the characterisation techniques mentioned above may give us some quantitative concept about the surface, more straightforward information can very often be obtained by using one's eyes. Visual inspection, therefore, plays an important and sometimes crucial role in 3-D topography analysis.[33] With the advent of fast computers with large memory capacities, its use is becoming more widespread. Commonly-used visual plots of surface topography are –

- axonometric projection,
- contour mapping,
- inversion,
- truncation, and
- grey-scale mapping.

With further development in image processing techniques and the use of these techniques in surface topography analysis, some more meaningful visual inspection pictures for representing the surface topography will be developed.

2.3.12.6 Fractal Characterisation

Most statistical parameters are scale-dependent and in some cases, they may not represent surface characteristics correctly, especially for very fine surfaces. However, it is recognised[34] that many engineering surfaces show fractal characteristics, and that some scale-independent parameters may be extracted by using fractal theory.

Fractals are defined as an infinitely large range of geometric structures, regular or random, that exhibit 'self-similarity' over all ranges of scale, i.e. the structures are similar at all scales and they are characterised by some dimension which needs not coincide with the Euclidean dimension of space within which the structures occur.[34] At present, some of fractal characterisation approaches[35–37] such as the structural function and the compass method have been proposed, but since the fractal parameters lack straight-forward physical meanings it will be some time yet before this technique is widely accepted in industry.

2.3.13 Parameter Rash?

The problem of parameter rash[9] which hindered positive progress towards the development of an integrated approach to 2-D surface topography is becoming visible in 3-D analysis. Statistical, fractal, spectral, time series, functional and other characterisation techniques are widely used, many of them generating parameters ranging from visual to fractal.

Parameters generated by users' 3-D systems were as follows:

(1) Parameters extended from their 2-D counterparts.

(2) Anisotropic index, anisotropic direction.

(3) Distribution and quantification of motifs.

(4) Bearing area ratio, contact area.

(5) Oil volume, volume of pile-up and indent.

(6) Geometric parameters, patterning parameters.

(7) Fractal dimensions.

(8) Averaged 2-D parameters on a surface.

(9) Slope and curvature distribution.

It is significant to note that as many as 62.5% of respondents admit that their adopted parameters do not give them all the information they would like to have about the geometry and functionality of the surface. This clearly shows that more effective and functionally significant characterisation methods are required.

2.4 CONCLUSIONS

The following general conclusions can be drawn from the survey:

- 3-D surface topography now spans a broad range of application industries – automotive, chemical, steel, machine tools.

- It was clear that there has been a remarkable shift in favour of optical instruments. There were more optical instrument users recorded in the survey, than stylus instrument users.

- Digital filtering techniques were used in almost all cases.

- Better measurement techniques are needed due to the limitations of some widely-used instruments in terms of scope, accuracy, environmental considerations and operation speed.

- The problem of parameter rash is becoming visible in 3-D analysis. Many researchers have introduced parameters, many of which are directly derived from 2-D ones. Statistical, fractal, spectral, time series, functional and other characterisation techniques are widely used, many of them generating parameters ranging from visual to fractal.

- It is also clear, that although significant 3-D surface topography analysis systems and characterisation approaches have been developed, there is a lack of a unified and comprehensive standard for data capture procedures, data analysis and result interpretation in terms of the surface geometry and hence the functional significance.

- Surface characterisation, representation and classification are not fully understood, and new approaches are required for the characterisation of different surfaces.

- It is essential to properly characterise the functional properties of surfaces in order to maximise their efficiency in any particular engineering application.

- There was a general consensus that functional parameters have not been properly investigated and that there was a requirement for 3-D parameters to supply functional information.

- Perhaps the most important factor is the pressing challenge for the development of an integrated theoretical basis for surface topography measurement, representation, and interpretation – perhaps a better definition of the surface geometry, from which structured and consistent deductions can be made in the light of a particular functional requirement.

Part III

VISUALISATION TECHNIQUES AND A PRIMARY PARAMETER SET FOR CHARACTERISING THREE-DIMENSIONAL SURFACE TOPOGRAPHY

W P Dong, N L Luo
P J Sullivan and K J Stout

VISUALISATION TECHNIQUES AND A PRIMARY PARAMETER SET FOR CHARACTERISING THREE- DIMENSIONAL SURFACE TOPOGRAPHY

In this Part visualisation techniques and a primary parameter set for characterising 3-D surface topography are introduced. The primary parameter set has been developed from the research results reported in the literature and carried out at the University of Birmingham recently. Statistical, geometrical and functional parameters as well as the visual plots are introduced. Parameter values calculated from some typical engineering surfaces are given.

NOMENCLATURE

$f(x,y)$	Reference datum.
h	Coordinate of normalised surface height.
M, N	The numbers of the sampling points in x and y directions respectively.
V_c	Void volume of the unit sampling area in the core zone.
V_v	Void volume of the unit sampling area in the valley zone.
x, y, z	Three dimensional coordinates.
$\Delta x, \Delta y$	Sampling intervals in x and y directions respectively.
$z(x,y)$	Original surface.
$\eta(x, y)$	Residual surface.
S_q	Root mean square deviation of the surface.
S_z	Ten point height of the surface.
S_{sk}	Skewness of surface height distribution.
S_{ku}	Kurtosis of surface height distribution.
S_{ds}	Density of summit of the surface.
S_{tr}	Texture aspect ratio of the surface.
S_{td}	Texture direction of the surface.
S_{al}	Fastest decay autocorrelation length.
$S_{\Delta q}$	Root-mean-square slope of the surface.
S_{sc}	Arithmetic mean summit curvature of the surface.
S_{dr}	Developed interfacial area ratio.
S_{bi}	Surface bearing index.
S_{ci}	Core fluid retention index.
S_{vi}	Valley fluid retention index.

3.1 INTRODUCTION

A key problem in surface-related research is choosing parameters that characterise surface properties in such a way that they correlate with surface formation mechanisms and functional behaviour in a fundamental way. A solution to the problem has been attempted by developing a range of characterisation techniques,[1,2] such as statistics, spectral analysis, autocorrelation analysis, time series modelling, fractals and function-specific methods. However, whatever technique is adopted, from an engineering point of view, the characteristics of surface topography are finally represented by the relevant parameters aided by several commonly-used visualisation techniques. In other words, the properties of surface topography are assessed in accordance with visual images and the parameter values of the surface. 2-D parameters and profile plots cannot provide adequate and reliable information for the analysis of intrinsically three dimensional surface topography, whereas 3-D parameters and 3-D graphics offer an attractive, dimension-opening, and realistic approach. It follows that the investigation and understanding of the algorithms and the intrinsic meanings of 3-D parameters is of critical importance to the assessment of surface topography.

Literature on the analysis and applications of 3-D surface topography is spread in a diverse range of disciplines, e.g. manufacturing, materials, biology, chemistry and medicine in the form of research papers[3–19] and product information. It is inconvenient and sometimes it may cause confusion for people who are new to this area but intend to use this approach to solve practical problems which cannot, perhaps, be solved successfully with 2-D techniques. Even for those who are familiar with conventional 2-D profile analysis, it still takes a great deal of effort to find and to understand 3-D parameters and assessment procedures and to accustom themselves to 3-D digital surface metrology.

In order to provide an integrated approach to the assessment of 3-D surface topography, and to provide a unified reference for 3-D parameters, this Part, first of all, introduces some visualisation techniques for visual characterisation and manipulation of 3-D surfaces. Then it presents a primary parameter set which was proposed in an EC workshop for 3-D surface characterisation[2] and later distributed widely in European industry and academia. The parameters involved in the set have been tested by characterising many different kinds of engineering surfaces. These parameters are necessary, though not complete, to characterise a surface. Full definitions and accessible algorithms are provided. The usefulness and effectiveness of these parameters and the visual plots in providing topographic information are established through the use of examples which were carried out on typical engineering surfaces.

3.2 SURFACE TOPOGRAPHY IN
THREE DIMENSIONS

The surface of a solid is that part which represents the boundaries between the solid body and its environment. Surfaces as physical entities possess many attributes, geometry being one of them. Surface geometry by nature is three-dimensional and the detailed features are termed topography. In many applications, topography represents the main external features of a surface.

Quantitative 3-D surface measurement invariably involves the use of digital computers in the measurement process due to the large amount of information involved. Computers are involved in various stages in the measurement process: control of data collection, data storage, processing, analysis and output of results. In a Euclidean co-ordinate system, a physical surface can be represented as a continous function $z(x,y)$ with two independent variables, x and y. By necessity, a physical surface $z(x,y)$ must be digitised in the measurement process in order to enable storage, processing and analysis of topographic information. Therefore in digital 3-D surface analysis we denote a discrete surface height by $z(x_i,y_j)$, and a discrete spatial position by $x_i = (i\text{-}1)\Delta x(i = 1,2,...,M), y_j = (j\text{-}1)\Delta y(j=1,2,...,N)$ where $\Delta x, \Delta y$ are the sampling intervals and M, N are the numbers of the sampling points in x and y directions respectively (Fig. 3.1).

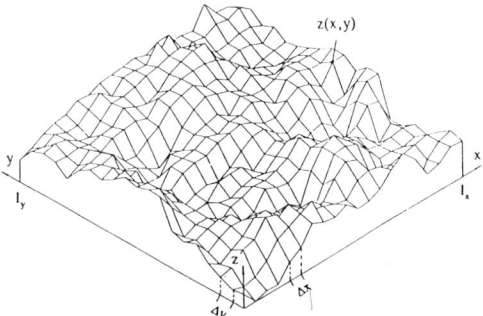

Fig. 3.1 Co-ordination of a digitised surface

3.3 REFERENCE DATUM FOR
TOPOGRAPHIC ANALYSIS

The reference datum used is the least squares mean plane which has the form of the geometrical surface and divides the surface such that, within the sampling area, the sum of squares of surface departures from this plane is minimised. A least squares mean plane may be defined by

$$f(x,y) = a + bx + cy \tag{3-1}$$

where a, b, c are the coefficients to be solved from the given topographic data. The sum of the squares of the surface departures from this plane is then given as

$$\varepsilon^2 = \sum_{l=1}^{N} \sum_{k=1}^{M} (z(x_k, y_l) - f(x_k, y_l))^2 = \sum_{l=1}^{N} \sum_{k=1}^{M} (z(x_k, y_l) - (a + bx_k + cy_l))^2 \tag{3-2}$$

The coefficients of the least squares mean plane are determined by minimising Equation (3–2) and are given as

$$b = \frac{12}{\Delta x} \cdot \frac{u - \dfrac{M-1}{2} w}{MN(M-1)(M+1)}$$

$$c = \frac{12}{\Delta y} \cdot \frac{v - \dfrac{N-1}{2} w}{MN(N-1)(N+1)} \tag{3-3}$$

$$a = \frac{(7MN + M + N - 5)w - 6(N+1)u - 6(M+1)v}{MN(M+1)(N+1)}$$

where

$$u = \sum_{l=1}^{N} \sum_{k=1}^{M} (k-1)z(x_k, y_l), \quad v = \sum_{l=1}^{N} \sum_{k=1}^{M} (l-1)z(x_k, y_l), \quad w = \sum_{l=1}^{N} \sum_{k=1}^{M} z(x_k, y_l) \tag{3-4}$$

The residual surface, $\eta(x,y)$ obtained by subtracting the least squares datum plane from the original surface, which is suitable for parameter evaluation is finally expressed as

$$\eta(x,y) = z(x,y) - (a + bx + cy) \tag{3-5}$$

3.4 VISUALISATION TECHNIQUES

The human brain still remains the most powerful and versatile intuitive processor of information, particularly when complicated and ill-defined but patterned data is involved. It is now generally accepted that a full intuitive appreciation of a surface can only be achieved by 3-D visualisation methods and that 2-D profiles are inadequate for the qualitative assessment of surfaces. The parameters to be introduced below give mostly quantitative characterisation; visual inspection, however, plays an important and sometimes crucial role in 3-D surface topography analysis and is far more important in 3-D surface analysis than in 2-D profile analysis. Sometimes it is the only objective in the use of 3-D methods. It can be stated that, to a great extent, the power of the 3-D approach rests largely on its visualisation capabilities. With the advent of fast computers with large memory capacities, this is becoming more widespread.[2,29,30] Sophisticated image processing and computer graphics techniques provide powerful and flexible tools for manipulating large amounts of topographic data.

Some useful and commonly used visual inspection methods and manipulation techniques in engineering applications are introduced below.

3.4.1 Visualisation Plots

The visualisation plots discussed here are mainly concerned with visual presentations of the original or processed topographic data. They do not include the presentation of mathematically restructured topographic information such as the height distribution, bearing area function, autocorrelation function and power spectrum. There are three basic types of visual presentations – isometric projection, contour mapping and greyscale image, each of which is based on a different graphic principle and exhibits different visual aspects of topographic data. Individual profiles can also be extracted from areal data and this can be from any specified direction and position where enough number of data points allows useful plotting.

3.4.1.1 Isometric Plot

An isometric plot is a projection of a 3-D object on a 2-D medium. An important property of this type of projection is that distances along the projection axes retain their original proportion.

The data points are interconnected with their neighbouring points by straight lines and hidden-line removal algorithms are usually used to avoid the display of invisible lines. The isometric projection gives a realistic view of the 3-D surface.

Two factors can be used to manipulate the visual property of the isometric plot. The first is the projection angle (from $0° - 90°$) which is an angle between the viewing direction and the X-Y plane. It allows emphasis of either amplitude (for smaller angles) or surface pattern features (for larger angles). Fig. 3.2 shows isometric plots of a faced turned surface with the projection angles $45°$ and $70°$.

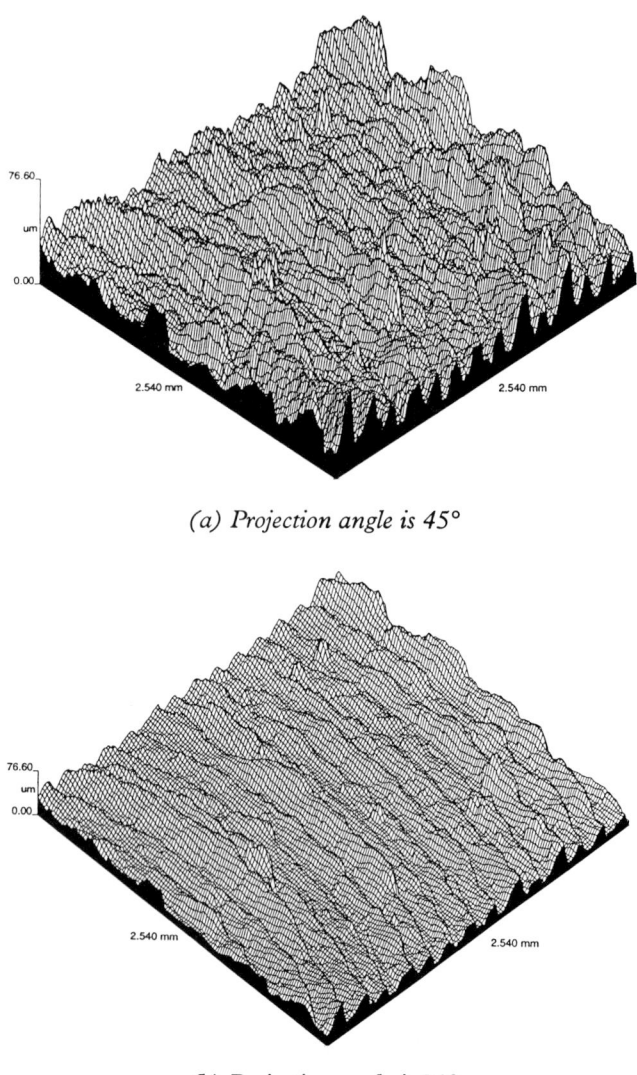

(a) Projection angle is 45°

(b) Projection angle is 70°

Fig. 3.2 Isometric plots of a face turned surface with different projection angles

The second factor is the rotation angle (from 0° – 360°) by which the X-Y plane rotates around the Z axis. It allows the isometric projection to expose different parts of the surface by choosing the parts of the surface that face the viewer. Fig. 3.3 shows the isometric plots of a bored surface with the rotation angles 0° and 90°.

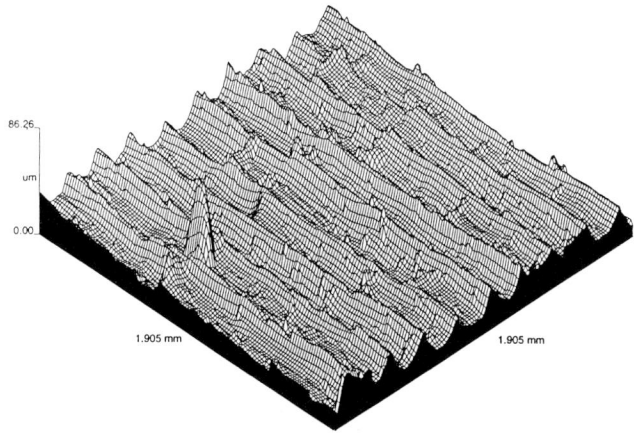

(a) Rotation angle is 0 °

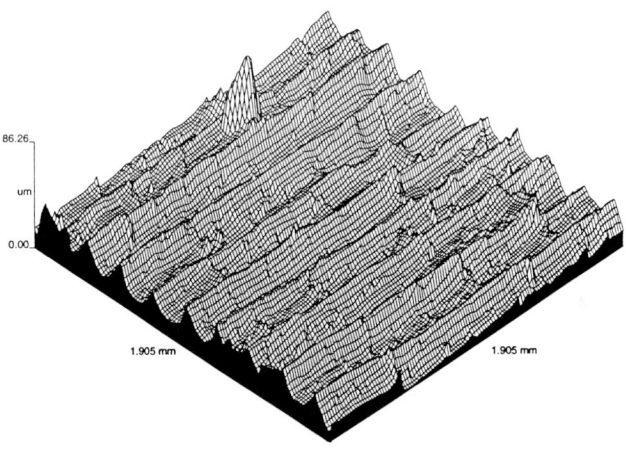

(b) Rotation angle is 90°

Fig. 3.3 Isometric plots of a bored surface with different rotation angles

3.4.1.2 Contour Plot

In a contour plot, lines are used to connect points of the same height in the same way as a terrain map is made. Fig. 3.4 shows the contour plots of a shaped surface. Except for the lines intersecting with the boundary of the mapped area, all lines are closed and do not intersect each other. The data points on the same line have the same height value. Although the contour plot is 2-D in essence, the shape of the contours, together with line numbering or colouring by heights, helps the viewer to reconstruct some 3-D information.

This type of presentation of topographic data is particularly useful for identifying directional features.

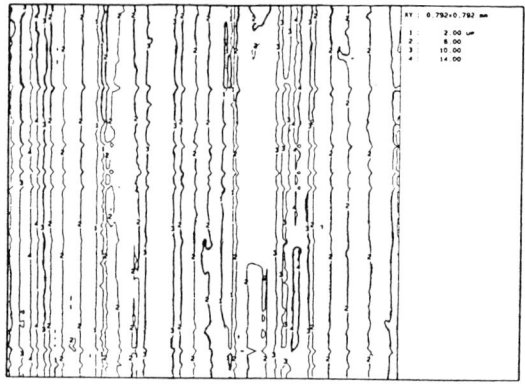

(a) The number of contour levels is 4

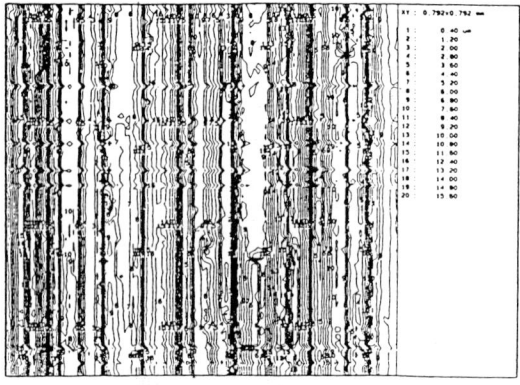

(b) The number of contour levels is 20

Fig. 3.4 Contour plots of a shaped surface

A factor to control the viewing property of the contour plot is the number of contour levels. A large number can reveal more detailed topographic features, but it may mask broad features. A small number of levels masks insignificant features and emphasises the dominant features. The commonly-used number of contour levels is between 4 and 40. The contour plots shown in Fig. 3.4 are drawn by 4 and 20 contour levels respectively.

3.4.1.3 Greyscale Image

In a greyscale image of a surface, the data points of the areal matrix are re-quantized to the resolution of greys available on the display device or media and displayed as greys proportional to the re-quantized surface height values. It gives a photographic image on the screen. The image quality is controlled by the number of grey levels. The useful number of grey levels ranges from 16 to 64 on a standard VGA monitor. Fig. 3.5 shows a greyscale plot of a honed surface with the grey levels 16 and 64.

(a) The number of grey levels is 16

(b) The number of grey levels is 64

Fig. 3.5 Intensity plots of a honed surface

3.4.2 Manipulation Techniques

The above-mentioned visual plots are different techniques for the visual presentation of topographic data. There are situations where topographic data needs to be manipulated in certain ways to aid visual or other interpretation. There are almost limitless applicable techniques to this effect, e.g., topography inversion, truncation, zoom or clipping, digital filtering, and a whole range of image processing and analysis techniques. It is important to be aware of the capabilities and limitations of these techniques and choose those that are useful for the intended purpose. Here we describe a few commonly used manipulation techniques. Their uses are not limited to visualisation.

3.4.2.1 Inversion

Inversion refers to the process of turning the surface upside down so that the pits, troughs and valleys become summits and ridges. This helps to reveal topographic features located at the lower part of the surface and is helpful for identifying significant feaatures in pits and troughs. Fig. 3.6 shows an isometric plot of an inverted ground surface; some pits and troughs can be clearly seen.

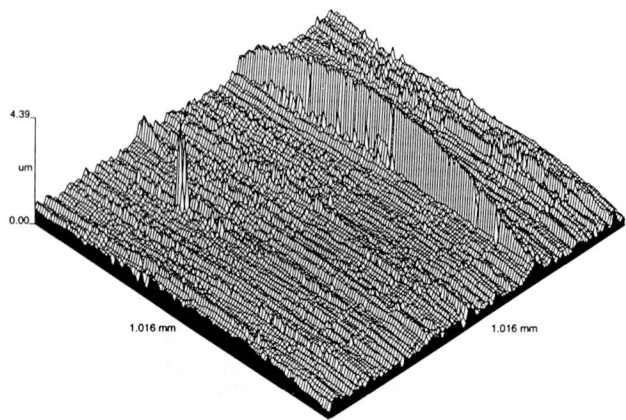

Fig. 3.6 Inversion of a ground surface

3.4.2.2 Truncation

Truncation refers to the removal of the material above a given plane that is parallel to the mean plane. Truncation simulates in simple terms the wear process and leaves the bearing area and contours visible when plotted, as shown in Fig. 3.7. It may also be used to reveal features below a certain level that are otherwise invisible. The truncation level can be specified either as a percentage value or an absolute value. An example is shown in Fig. 3.7 where a 40% and a 60% height truncation relative to the highest peak of an electrical discharge machining (EDM) surface are displayed.

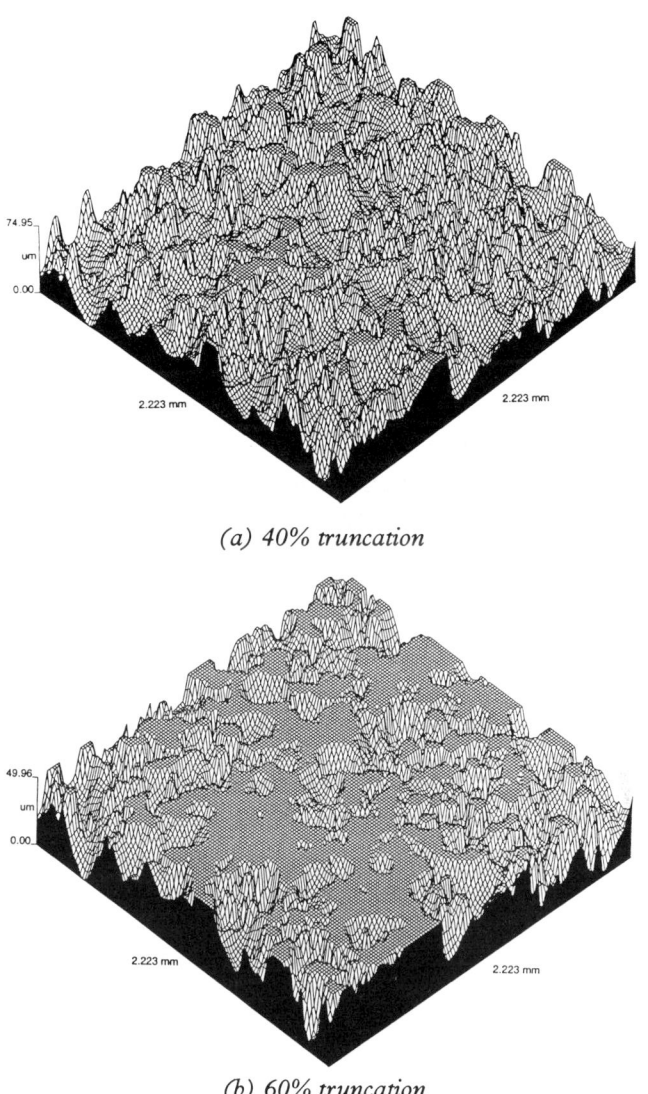

(a) 40% truncation

(b) 60% truncation

Fig. 3.7 Truncation of an EDM surface

The combination of inversion and truncation may allow the upper part of a surface to be reserved and the lower part of the surface to be removed as seen in Fig. 3.8 which shows the top 40% material of the milled surface shown previously in Fig. 3.7.

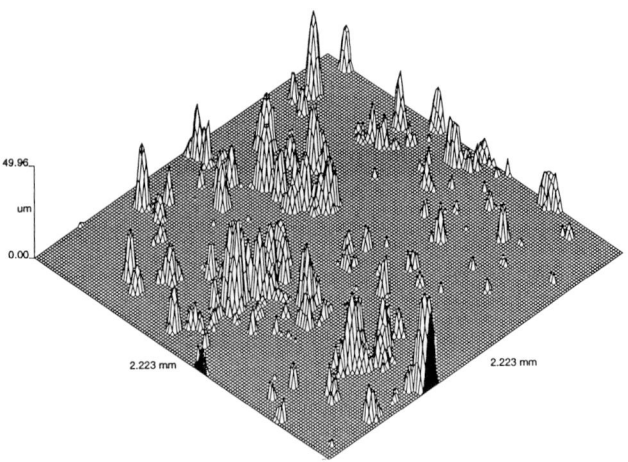

Fig. 3.8 Top 40% material display of the EDM surface shown in Fig. 3.7

3.4.2.3 Zooming and Clipping

It is sometimes necessary, in both visual representation and data analysis, to extract a sub-area from the original area in order to either study the isolated area in detail (*zoom*) or to remove unwanted data points within the isolated area (*clipping*). On a computer screen, the zooming process is often done by the selection of the area of interest with the cursor (Fig. 3.9(a)) and re-display the data of the small area on the screen in an enlarged size (Fig. 3.9(b)). Subsequent data manipulation, visualisation and analysis can be performed on the chosen area. However, the clipping process is used to select the unwanted area and remove all the data points inside the area. Subsequent data manipulation is done without including the removed data. Fig. 3.10 shows a contour map of a honed surface from which the dark area is clipped.

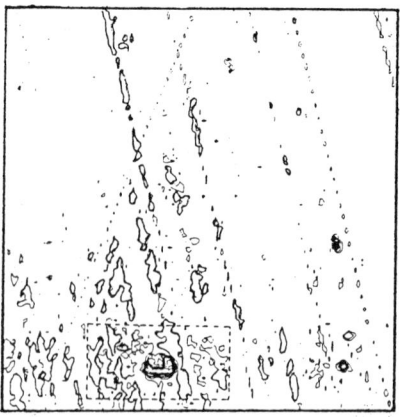

(a) Selection of the interested area

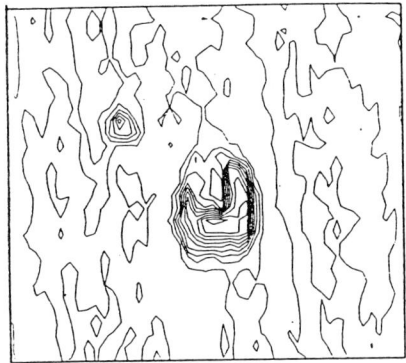

(b) Re-display the data of the interested area

Fig. 3.9 Illustration of the zoom technique

Fig. 3.10 Contour plot of a honed surface which is clipped in the dark area

3.4.2.4 Surface Image Enhancement

There are a host of image processing analysis techniques that can be used for the enhancement of visual presentations and feature extraction. An example of the techniques in this category is the use of edge-highlighting methods to emphasise visual features of a greyscale image. Derivatives of the first order or the second order of the data can be taken for this purpose. To emphasise features in a single direction, directional derivatives can be used. Fig. 3.11 presents greyscale images enhanced by plotting slopes in either x or y or both directions.

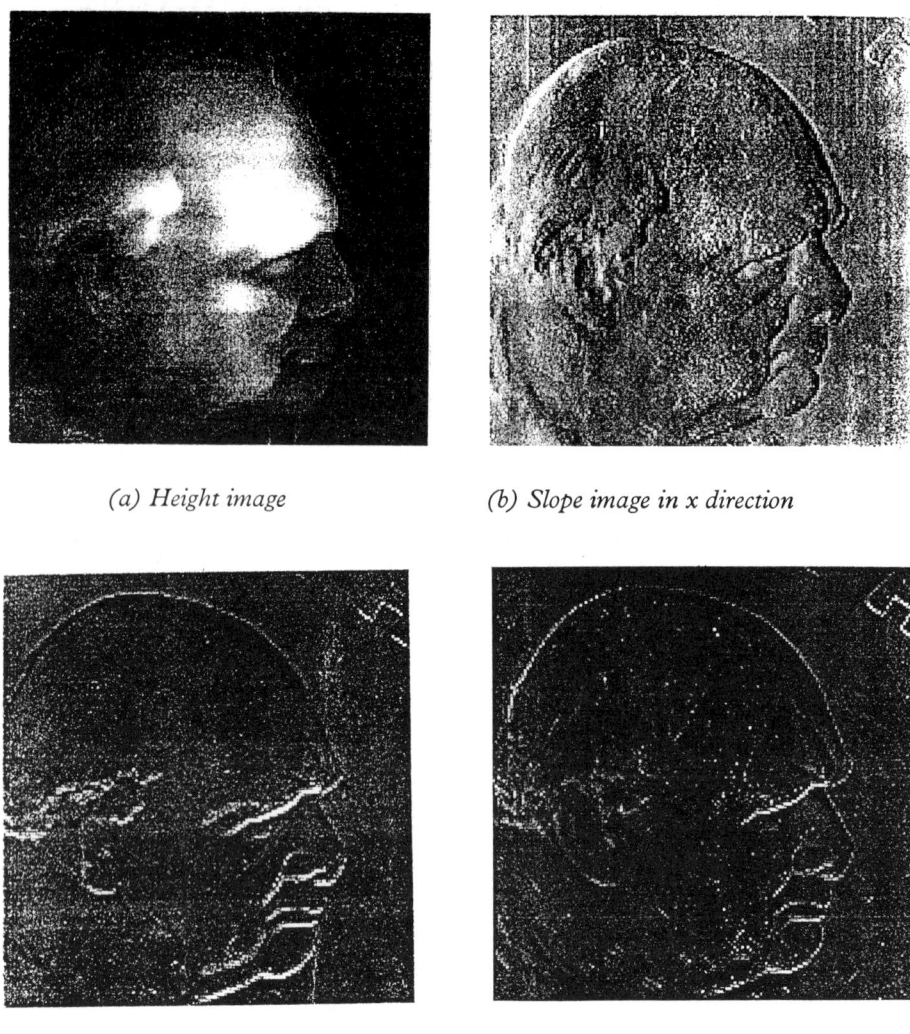

(a) Height image

(b) Slope image in x direction

(c) Slope image in y direction

(d) Slope image in both x & y directions

Fig. 3.11 Image enhancement

3.5 SPECIFICATION OF THE PARAMETERS

Before the 3-D parameters are given, some consideration of the conditions under which the parameters are defined and the experiments are carried out have to be specified.

(1) In order to distinguish between 2-D and 3-D parameters, different names are given to the 3-D parameters. The authors followed the general guidelines agreed among industrial users and surface metrologists who attended the workshop on the Characterisation of Surfaces in 3-D specially organised by the Commission of the European Communities, and which was held in Brussels in September 1991.[20] It was proposed at the workshop that a headed letter 'S' (for 'surface') should be used in 3-D instead of the letter 'R' in 2-D. Therefore, all 3-D parameters presented here are denoted by 'S' with appropriate subscripts.

(2) The reference plane used for characterising surfaces is the least square plane. Therefore the residual surface (equation (3–5)), which has a zero mean, is used in the algorithms.

(3) Unlike 2-D parameters defined in the national and international standards,[21–23] which are evaluated over several sampling lengths, 3-D parameters presented here are determined within one sampling area.

(4) For establishing the concept of how useful the proposed parameters are for some typical engineering surfaces, experiments were carried out to test the significance of the parameters. The surfaces tested are honed, plateau honed, ground, electro-discharge machined (EDM), bored, shaped, turned and milled. The size of the sampling matrix used here is 128x128 for all measurements; the sampling intervals were chosen in accordance with the texture of the surfaces.

3.6 A PRIMARY PARAMETER SET

The primary parameter set proposed in the second Workshop on the Characterisation of Surfaces in 3-D[2] involves 14 parameters which characterise some major aspects of topographic features. There are four parameters for describing the amplitude and height distribution properties, three parameters for describing spatial properties, three parameters for describing hybrid (i.e. both amplitude and spatial) properties and three parameters for some functional properties.

3.6.1 Amplitude and Height Distribution Parameters

The parameters defined in this category are extensions of corresponding 2-D parameters. Two of them are defined to describe the amplitude property and two of them are defined to describe the shape of the surface height distribution.

3.6.1.1 Root-Mean-Square Deviation S_q

This is the root mean square (RMS) value of the surface departures within the sampling area (R_q for 2-D).[21–23] The digital formula is

$$S_q = \sqrt{\frac{1}{MN} \sum_{j=1}^{N} \sum_{i=1}^{M} \eta^2(x_i, y_j)} \tag{3–6}$$

This is a widely used parameter which indicates surface roughness in a well known statistical form. It is insensitive to the sampling intervals adopted, but is sensitive to the size of the sampling area and the frequency band if a 2-D filter is used.

3.6.1.2 Ten Point Height of the Surface S_z

This is an extreme parameter defined as the average value of the absolute heights of five highest peaks and the depths of five deepest pits or valleys within the sampling area (R_z for 2-D).[21–23] In digital form, the formula for S_z is

$$S_z = \frac{\sum_{i=1}^{5} |\eta_\pi| + \sum_{i=1}^{5} |\eta_{vi}|}{5} \tag{3–7}$$

where η_{pi} and η_{vi} ($i = 1, 2, ..., 5$) are the five highest surface summits and lowest surface valleys respectively.

A problem arises when calculating the parameter with digital computers, i.e. the definition of summits and valleys in areal topographic data. They are more ambiguous compared with the definition of the peaks and valleys of profile data. Firstly, a summit may be defined variously to lie within four nearest neighbours, eight nearest neighbours,[24] a small zone summit[25] and/or contour-based summit.[26] By using different definitions, the number of summits and the average height of the summits would be different. Secondly, the ridges, saddles, valleys and local undulations around the highest summit may adversely affect the ability to find the second and subsequent highest summits. Therefore the ten point height S_z is a summit definition dependent parameter. Whenever the parameter is given, the definition of the summit should be specified. Generally speaking, the ten point height is more sensitive to the adoption of the sampling interval than the RMS deviation. Fig. 3.12 is a plot of the ten point summit, S_z, based on eight nearest neighbours versus the RMS deviation, S_q.

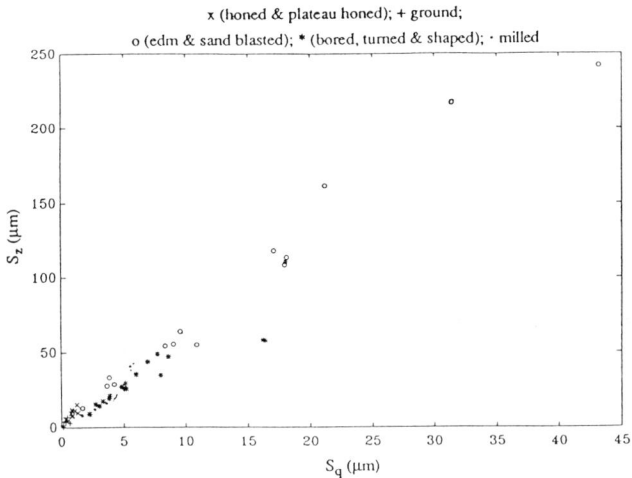

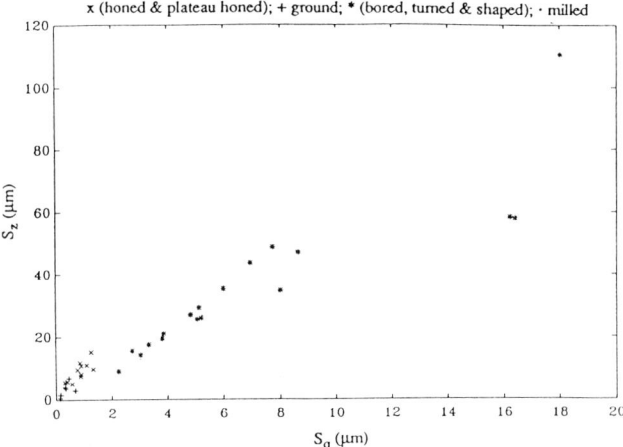

Fig. 3.12 Diagrams of S_z versus S_q for typical engineering surfaces

3.6.1.3 Skewness of Surface Height Distribution S_{sk}

This is the measure of asymmetry of surface deviations about the mean plane (S_k for 2-D).[22,23] It is given by the formula

$$S_{sk} = \frac{1}{MNS_q^3} \sum_{j=1}^{N} \sum_{i=1}^{M} \eta^3(x_i, y_j) \tag{3-8}$$

This parameter can be effectively used to describe the shape of the surface height distribution. For a Gaussian surface which has a symmetric surface height distribution, the skewness is zero. For an asymmetric surface height distribution, the skewness may be negative if the distribution has a longer tail at the lower side of the mean plane or positive if the distribution has a longer tail at the upper side of the mean plane. The influence of the sampling interval on the parameter is slightly more significant than is found in the RMS deviation.

3.6.1.4 Kurtosis of Surface Height Distribution S_{ku}

This is a measure of the peakedness or sharpness of the surface height distribution (kurtosis for 2-D).[22] It is given by the formula

$$S_{ku} = \frac{1}{MNS_q^4} \sum_{j=1}^{N} \sum_{i=1}^{M} \eta^4(x_i, y_j) \tag{3-9}$$

This parameter characterises the spread of the height distribution. A Gaussian surface has a kurtosis value of 3. A centrally distributed surface has a kurtosis value larger than 3 whereas the kurtosis of a well spread distribution is smaller than 3. The influence of the sampling interval on the parameter is again slightly more significant than is found in the RMS deviation.

A diagram showing S_{ku} versus S_{sk} for typical engineering surfaces is given in Fig. 3.13.

3.6.2 Spatial Parameters

Four parameters are proposed to characterise the following four spatial properties.

- Density of summits.
- Surface texture aspect ratio.
- Directionality of surface lay.
- The fastest decay of the 2-D autocorrelation function.

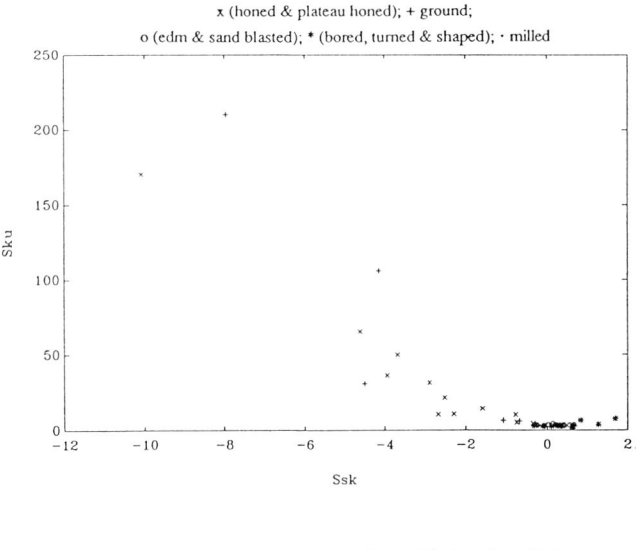

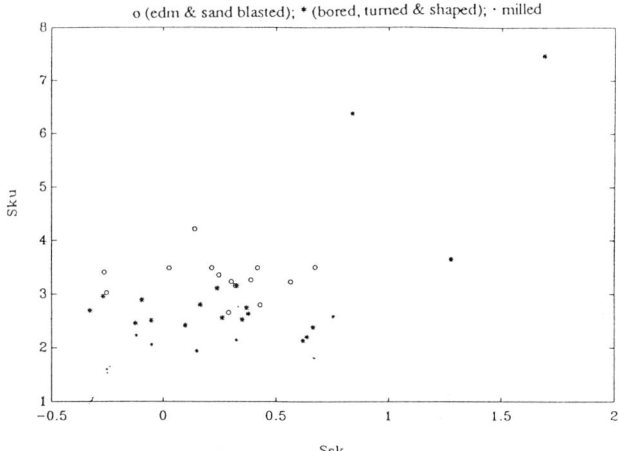

Fig. 3.13 Diagrams of S_{ku} versus S_{sk} for typical engineering surfaces

3.6.2.1 Density of Summits of the Surface S_{ds}

This is the number of summits of a unit sampling area. It is given by the formula,

$$S_{ds} = \frac{\text{Number of summits}}{(M-1)(N-1)\cdot\Delta x\cdot\Delta y} \qquad (3\text{--}10)$$

As in the determination of S_z, S_{ds} varies with the definition of the summit. For different definitions of the summit, the density of the summit would be quite different. Therefore, whenever this parameter is used, the definition of the summit used has to be specified. Moreover, this parameter would be severely influenced by the adoption of the sampling interval in measurement. For reference, the density of summit (based on eight nearest neighbours) of surfaces shown in Figs. 3.2–7 is listed in Table 3.1.

Table 3.1 Parameters of some surfaces

Surfaces	Turned Fig 3.2	Bored Fig 3.3	Shaped Fig 3.4	Honed Fig 3.5	Ground Fig 3.6	Milled Fig 3.7
S_{ds} (1/mm^2)	68.67	87.08	42.01	433.4	1119	68.63
S_{tr}	0.2	0.062	0.075	0.162	0.047	0.87
S_{al} (μm)	60	60	102	28	24	123
$S_{\Delta q}$ (μm/μm)	0.427	0.272	0.38	0.09	0.04	0.587
S_{sc} (1/μm)	0.023	0.028	0.024	0.0083	0.0045	0.046

3.6.2.2 Texture Aspect Ratio of the Surface S_{tr}

This parameter is a measure of surface texture (*long-crestedness* or *isotropy*). It is defined on the areal autocorrelation function (AACF, see Fig. 3.14(b)-3.16(b)), i.e.

$$0 < S_{tr} = \frac{\text{The fastest decay distance to 0.2 on the normalised AACF}}{\text{The slowest decay distance to 0.2 on the normalised AACF}} \leq 1 \quad (3\text{-}11)$$

This parameter is used to identify texture pattern i.e. isotropy or anisotropy. In principle, the texture aspect ratio would be in the range (0,1). Larger values (S_{tr}>0.5) of the ratio indicates stronger isotropy, e.g. an EDM surface shown in Fig. 3.14 whose S_{tr}=0.83, whereas smaller values (S_{tr}<0.3) of the ratio indicates stronger anisotropy, e.g. a shaped surface shown in Fig. 3.15 whose S_{tr}=0.075. For some surfaces, which can be regarded as neither significantly layed nor isotropic, their texture aspect ratio values are between those of strongly layed and strongly isotropic surfaces. A honed surface usually possesses a significant crossed lay texture (Fig. 3.16 whose S_{tr}=0.063); its texture aspect ratio is smaller than 0.3. A change of the texture aspect ratio of a surface from less than 0.3 to larger than 0.5 may indicate that the texture has been worn significantly. The values of S_{tr} of surfaces shown in Figs. 3.2–7 are given in Table 3.1. The influence of the sampling interval on the parameter is negligible.

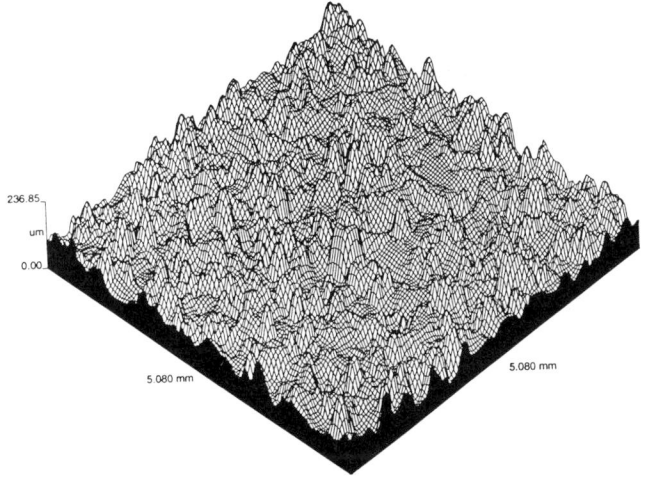

(a) Isometric plot

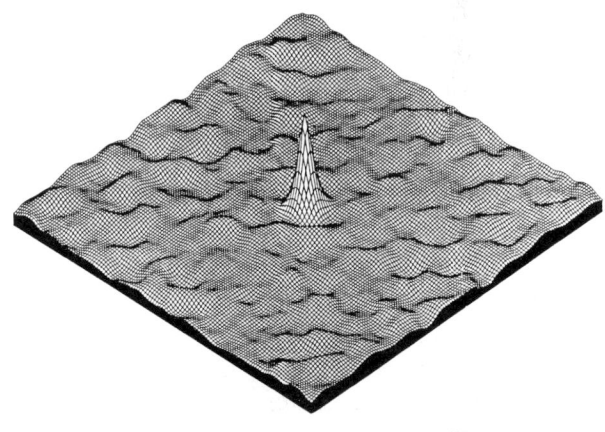

(b) AACF

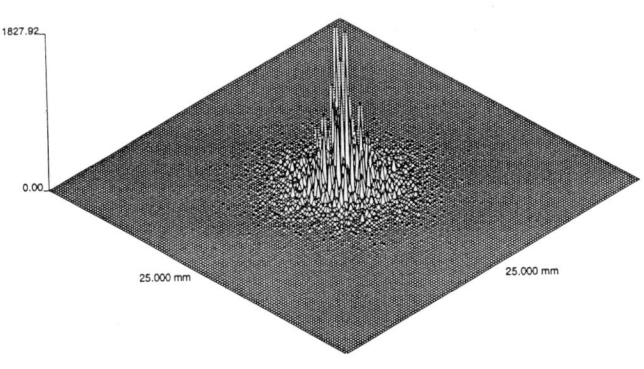

(c) APSD

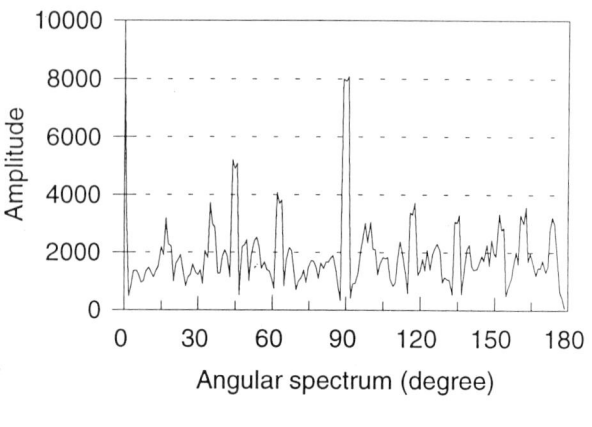

(d) Angular spectrum

Fig. 3.14 Isotropic topography of an EDM surface

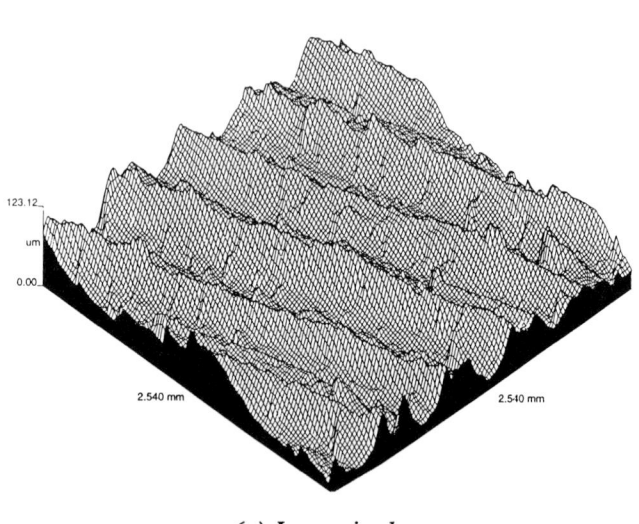

(a) Isometric plot

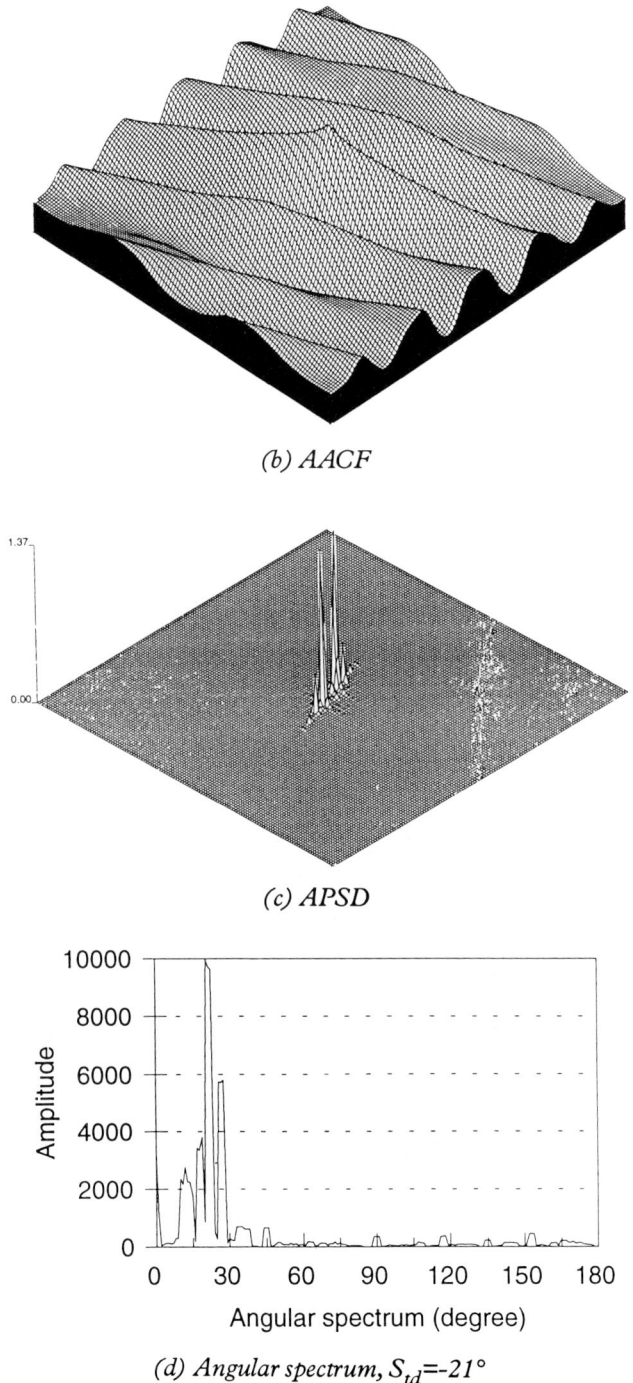

(b) AACF

(c) APSD

(d) Angular spectrum, S_{td}=-21°

Fig. 3.15 Linear texture of a shaped surface

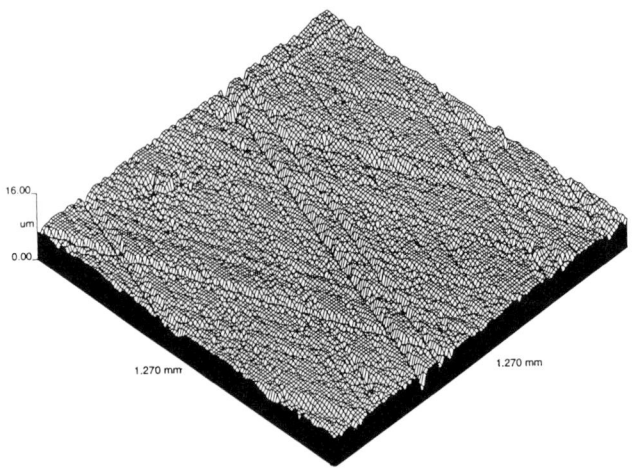

(a) Isometric plot

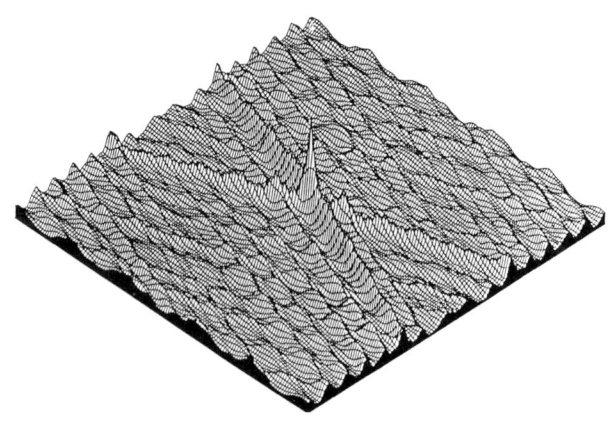

(b) AACF

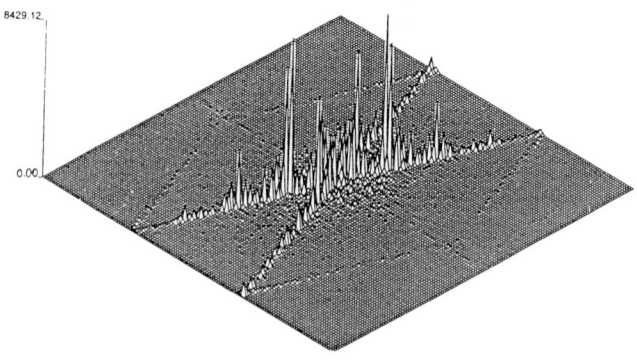

(c) APSD

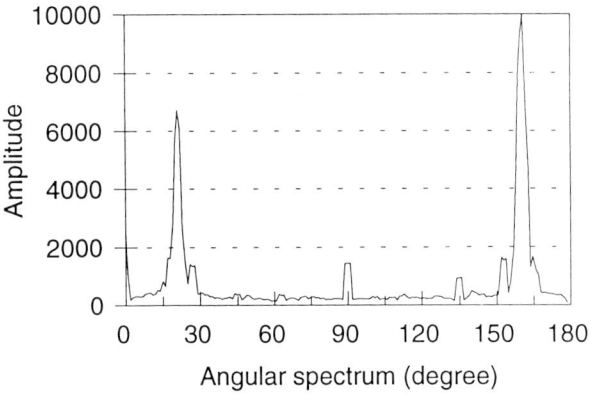

(d) Angular spectrum, $S_{td}=19°$ and $S_{td}=-21°$

Fig. 3.16 Crossed lay texture of a plateau honed surface

3.6.2.3 Texture Direction of the Surface S_{td}

This is the parameter used to give the pronounced direction of the surface texture with respect to the y axis, i.e. it gives the lay direction of the surface. For a unified definition of the texture direction, a surface texture shown in the right hand side of Fig. 3.17 is given a positive angle, whilst a surface texture shown in the left hand side of Fig. 3.17 is given a negative angle. By this definition, when the measurement trace direction is perpendicular to the lay (this is a recommended practice) the texture direction is 0°.

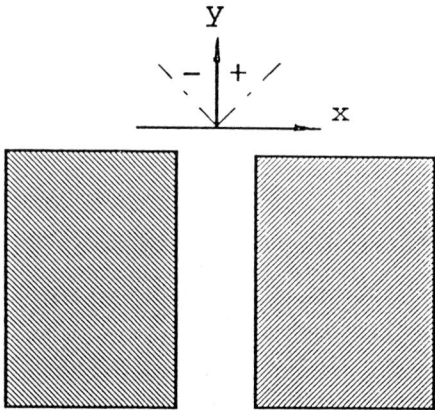

Fig. 3.17 Definition of the texture direction

An effective way to obtain the texture direction is to calculate areal power spectral density (APSD) of the surface, and then to determine the position where the maximum power appears in the angular spectrum.[2] Fig. 3.15(c) is an APSD of a shaped surface (Fig. 3.15(a)), Fig. 3.15(d) is its angular spectrum. The angle where the maximum power appears is termed β. Therefore the texture direction is determined by

$$S_{td} \begin{cases} -\beta \, , & \beta \le \dfrac{\pi}{2} \\ \pi - \beta \, , & \dfrac{\pi}{2} < \beta \le \pi \end{cases} \tag{3–12}$$

Thus the texture direction of the surface shown in Fig. 3.15 is S_{td}=-21°. The application of the texture direction is conditional. It would be meaningless if the texture aspect ratio, S_{tr}, is larger than, say S_{tr}>0.5. As is seen in Fig. 3.14 whose S_{tr}=0.83, no significant texture is observed. For curved texture, e.g. a face turned surface, the texture direction is then a general tangential direction of the curvature within the mapped area. For crossed lay surfaces, e.g. a honed surface shown in Fig. 3.16, two crossed directions might be determined from the angular spectrum by finding the positions where the maximum and the second maximum peaks are located (S_{td} = 19° and S_{td} = -21°).

3.6.2.4 Fastest decay autocorrelation length S_{al}

This is a parameter in length dimension used to describe the autocorrelation character of the AACF. It is defined as the horizontal distance of the AACF which has a fastest decay to 0.2. In other words the S_{al} is the shortest autocorrelation length during which the AACF decays to 0.2 in any possible direction. Mathematically it is given by

$$S_{al} = \min \left(\sqrt{\tau_x^2 + \tau_y^2} \right) , \quad \tilde{R}\,(\tau_x, \tau_y) \le 0.2 \tag{3–13}$$

where $\tilde{R}\,(\tau_x, \tau_y)$ is the normalised AACF, τ_x and τ_y are the coordinates of the AACF in x and y directions respectively. For an anisotropic surface S_{al} is obtained in the direction perpendicular to the surface lay. A large value of the S_{al} denotes that the surface is dominated by low frequency (or long wavelength) components, while a small value of the S_{al} denotes an opposite situation. The values of S_{al} of surfaces shown in Figs. 3.2–7 are given in Table 3.1. The influence of the sampling interval on the parameter is, in most cases, negligible.

3.6.3 Hybrid Parameter

The hybrid property is a combination of amplitude and wavelength. Any changes which occurred in either amplitude or space may have effects on the hybrid property. Three hybrid parameters are presented here.

3.6.3.1 Root-Mean-Square Slope of the Surface $S_{\Delta q}$

This is the root-mean-square slope of the surface within the sampling area (Δ_q for 2-D).[23] The surface slope at any point is represented by[5,27]

$$\rho_{ij} = \left[\left(\frac{\partial \eta(x,y)}{\partial x} \right)^2 + \left(\frac{\partial \eta(x,y)}{\partial y} \right)^2 \right]^{1/2} \Bigg|_{x=x_i, y=y_j}$$

$$\approx \left[\left(\frac{\eta(x_i,y_j) - \eta(x_{i-1},y_j)}{\Delta x} \right)^2 + \left(\frac{\eta(x_i,y_j) - \eta(x_i,y_{j-1})}{\Delta y} \right)^2 \right]^{1/2}$$

(3–14)

thus, the root-mean-square slope is given by

$$S_{\Delta q} = \sqrt{\frac{1}{(M-1)(N-1)} \sum_{j=2}^{N} \sum_{i=2}^{M} \rho_{ij}^2} = \sqrt{\frac{1}{(M-1)(N-1)} \sum_{j=2}^{N} \sum_{i=2}^{M} \left[\left(\frac{\eta(x_i,y_j) - \eta(x_{i-1},y_j)}{\Delta x} \right)^2 + \left(\frac{\eta(x_i,y_j) - \eta(x_i,y_{j-1})}{\Delta y} \right)^2 \right]}$$

(3–15)

The root-mean-square slope of the surface is sensitive to the sampling interval. The values of the root-mean-mean slope of some engineered surfaces shown in Figs. 3.2–7 are listed in Table 3.1.

3.6.3.2 Arithmetic Mean Summit Curvature of the Surface S_{sc}

This is defined as an average of the principal curvatures of the summits within the sampling area. Since the sum of the curvatures of a surface at a point along any two orthogonal directions is equal to the sum of the principal curvatures,[5] the arithmetic mean summit curvature of the surface is then given by

$$S_{sc} = -\frac{1}{2} \cdot \frac{1}{n} \sum_{k=1}^{n} \left(\frac{\partial \eta^2(x,y)}{\partial x^2} + \frac{\partial \eta^2(x,y)}{\partial y^2} \right) \Bigg|_{\text{for any summit}}$$

$$\approx -\frac{1}{2} \cdot \frac{1}{n} \sum_{k=1}^{n} \left(\frac{\eta(x_{p+1},y_q) + \eta(x_{p-1},y_q) - 2\eta(x_p,y_q)}{\Delta x^2} + \frac{\eta(x_p,y_{q+1}) + \eta(x_p,y_{q-1}) - 2\eta(x_p,y_q)}{\Delta y^2} \right),$$

for any summit located at x_p and y_q

(3–16)

This parameter can only be calculated after the summits (which rely on an accepted definition) have been found. It is sensitive to the sampling interval as well. The values of the arithmetic mean summit of some engineered surfaces shown in Figs. 3.2–7 are given in Table 3.1.

3.6.3.3 Developed Interfacial Area Ratio S_{dr}

This is the ratio of the increment of the interfacial area of a surface over the sampling area. The pile-up element, i.e. the smallest sampling quadrilateral ABCD[12] at (x_i, y_j) (i=1, 2, ..., M-1; j=1, 2, ..., N-1), is shown in Fig. 3.18. Thus the interfacial area of the quadrilateral is defined as an average of two sets of triangle areas (ABC & ACD and ABD & BCD) given by,

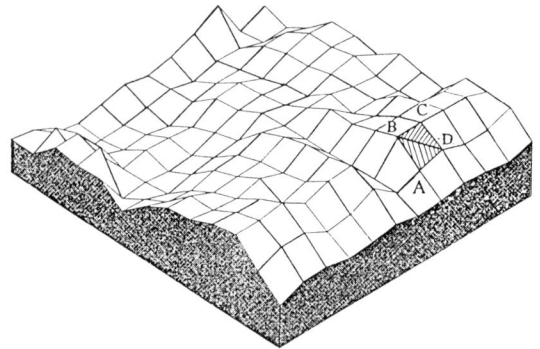

Fig. 3.18 Schematic diagram of the interfacial area

$$A_{ij} = \frac{1}{2}\left[\left(\frac{1}{2}\left|\overrightarrow{BA}\times\overrightarrow{BC}\right| + \frac{1}{2}\left|\overrightarrow{DA}\times\overrightarrow{DC}\right|\right) + \left(\frac{1}{2}\left|\overrightarrow{AB}\times\overrightarrow{AD}\right| + \frac{1}{2}\left|\overrightarrow{CB}\times\overrightarrow{CD}\right|\right)\right] = \frac{1}{4}\left(\left|\overrightarrow{AB}\right|+\left|\overrightarrow{CD}\right|\right)\left(\left|\overrightarrow{AD}\right|+\left|\overrightarrow{BC}\right|\right)$$

$$= \frac{1}{4}\left\{\left(\left[\Delta y^2 + \left(\eta(x_i,y_j)-\eta(x_i,y_{j+1})\right)^2\right]^{1/2} + \left[\Delta y^2 + \left(\eta(x_{i+1},y_{j+1})-\eta(x_{i+1},y_j)\right)^2\right]^{1/2}\right).$$

$$\left(\left[\Delta x^2 + \left(\eta(x_i,y_j)-\eta(x_{i+1},y_j)\right)^2\right]^{1/2} + \left[\Delta x^2 + \left(\eta(x_i,y_{j+1})-\eta(x_{i+1},y_{j+1})\right)^2\right]^{1/2}\right)\right\}$$

$$(3-17)$$

The total developed area on the surface is

$$A = \sum_{j=1}^{N-1}\sum_{i=1}^{M-1} A_{ij}$$

$$(3-18)$$

Therefore the developed interfacial area ratio is given by

$$S_{dr} = \frac{\sum\limits_{j=1}^{N-1} \sum\limits_{i=1}^{M-1} A_{ij} - (M-1)(N-1)\Delta x \cdot \Delta y}{(M-1)(N-1)\Delta x \cdot \Delta y} \cdot 100\% \qquad (3\text{--}19)$$

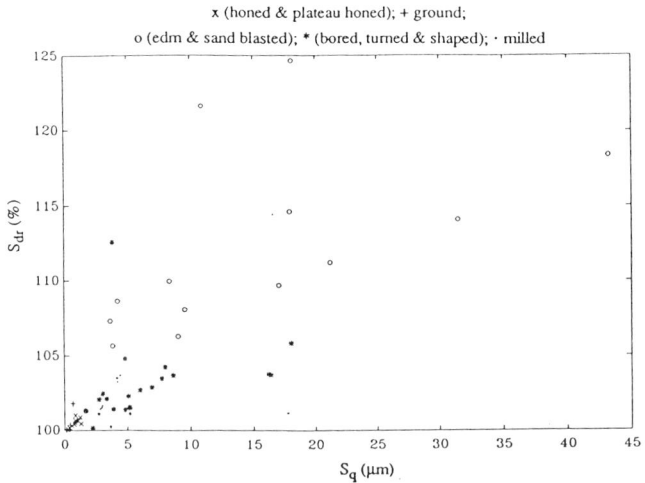

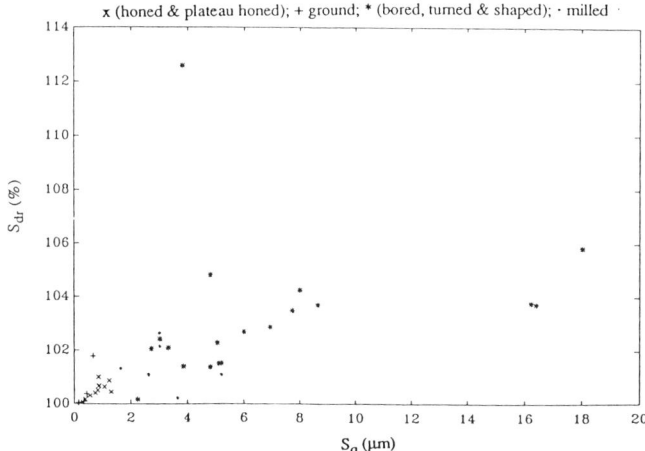

Fig. 3.19 Diagrams of the developed area ratio S_{dr} versus S_q of typical engineering surfaces

The developed interfacial area ratio reflects the hybrid property of surfaces. A large value of the parameter indicates the significance of either amplitude or spacing or both. This parameter is sensitive to the sampling interval. A diagram of the developed interfacial area ratio versus the root-mean-square height of engineering surfaces is given in Fig. 3.19.

3.6.4 Functional Parameters for Characterising Bearing and Fluid Retention Properties

There are a variety of functional applications of surfaces. It is impossible to define a functional parameter set to cover whole areas of functional applications. Therefore, the definitions of functional parameters would only be concentrated on some important and frequently applied aspects. The three functional parameters presented here are mainly defined for characterising surface bearing, fluid retention and relevant properties.

Area and volume are intrinsic geometric properties of surfaces; they are also related to surface bearing and fluid retention. Fig. 3.20 is a bearing area ratio curve established with areal topographic data. This curve is normalised to the RMS value, i.e. the vertical coordinate is denoted by

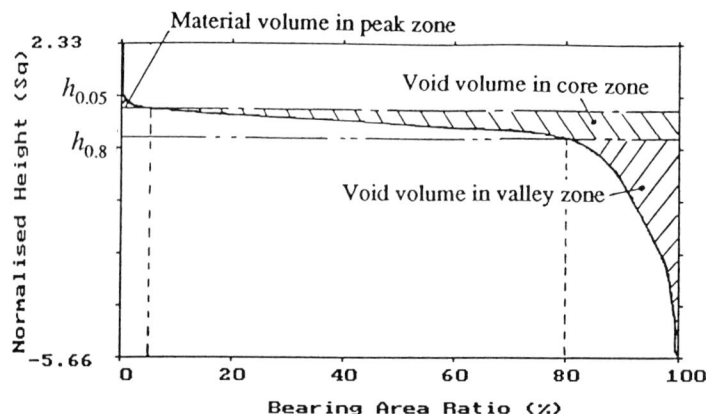

Fig. 3.20 A bearing area ratio curve scaled according to the RMS deviation

$$h = \eta(x,y)/S_q. \tag{3-20}$$

Similar to the analysis of R_k parameters standardised in DIN 4776,[28] the surface bearing area ratio is divided into three zones. The peak zone is between two horizontal lines, i.e. the top line (0% bearing area) and the 5% bearing area line (see Fig. 3.20). The core zone is between 5% and 80% bearing lines and the valley zone is between the 80% bearing line and the bottom line

(100% bearing area). It should be emphasised that the material volume of the unit sampling area at each zone is the area between the bottom line of the zone and the surface bearing area ratio curve, whilst the void volume of the unit sampling area at each zone is the area between the top line of the zone and the surface bearing area ratio curve.

In considering the scale of roughness, three indexes are defined to characterise bearing and fluid retention properties.

3.6.4.1 Surface Bearing Index S_{bi}

This is the ratio of the RMS deviation over the surface height at 5% bearing area, i.e.

$$S_{bi} = \frac{s_q}{\eta_{0.05}} = \frac{1}{h_{0.05}} \qquad (3\text{--}21)$$

where $\eta_{0.05}$ is the surface height at 5% bearing area. A larger surface bearing index indicates a good bearing property. For a Gaussian surface, the surface bearing index is about 0.608.

3.6.4.2 Core Fluid Retention Index S_{ci}

This is the ratio of the void volume of the unit sampling area at the core zone over the RMS deviation, i.e.

$$S_{ci} = \frac{V_c}{S_q} \qquad (3\text{--}22)$$

where V_c is the void volume of the unit sampling area at core zone. A larger S_{ci} indicates a good fluid retention in this zone. For a Gaussian surface, this index is approximately 1.56.

3.6.4.3 Valley Fluid Retention Index S_{vi}

This is the ratio of the void volume of the unit sampling area at the valley zone over the RMS deviation, i.e.

$$S_{vi} = \frac{V_v}{S_q} \qquad (23)$$

where V_v is the void volume of the unit sampling area at core zone. A larger S_{vi} indicates a good fluid retention in the valley zone. For a Gaussian surface, this index is about 0.11.

Diagrams of the valley fluid retention index versus the surface bearing index and versus the core fluid retention index of some engineering surfaces are shown in Fig. 3.21 and Fig. 3.22 respectively. There is a distinctive separation of functionally different surfaces manufactured by different processes.

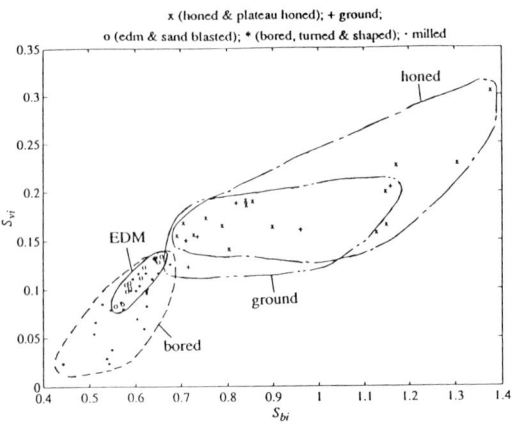

Fig. 3.21 A diagram of the valley fluid retention index S_{vi} versus the surface bearing index S_{bi} of typical engineering surfaces

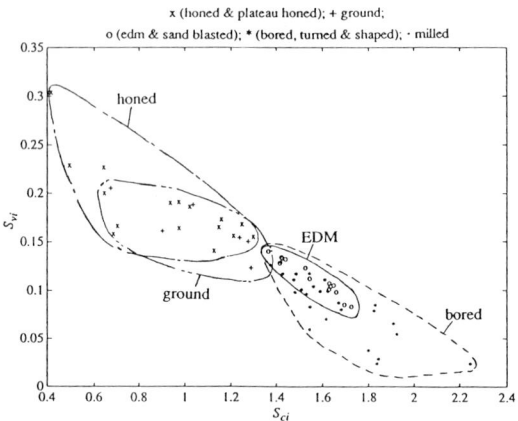

Fig. 3.22 A diagram of the valley oil retention index S_{vi} versus the core fluid retention index S_{ci} of typical engineering surfaces

3.7. CONCLUSIONS

A primary parameter set and visualisation techniques for characterising 3-D surface topography have been presented. Amplitude, spatial, hybrid and functional parameters, visual plots and manipulation techniques have been discussed. Their definitions and algorithms have been specified. Some experimental results have been given to establish the concept of the parameters and visualisation techniques. The authors believe that it represents a comprehensive, though not complete, analysis for 3-D surface topography, and encourage researchers and users to adopt the parameters and visualisation techniques introduced here.

Part IV

APPLICATIONS OF THREE-DIMENSIONAL SURFACE METROLOGY

W P Dong, L Blunt,
K J Stout and P J Sullivan

APPLICATIONS OF THREE-DIMENSIONAL SURFACE METROLOGY

In order to be able to present an overall view of the potential applications of three-dimensional surface metrology, this Part introduces some examples of practical uses where recently developed three-dimensional surface measurement and characterisation techniques have been used. All applications were carried out at the Centre for Metrology at the University of Birmingham, employing a comprehensive range of instruments including stylus, optical focus detection, optical interferometry and the scanning tunnelling microscope. These applications cover not only conventional engineering areas, but also non-traditional areas such as bio-engineering and physics. The performances of the instruments are discussed in terms of the above examples. The system set-ups are illustrated through the use of photographs of the instruments in use. A number of 3-D characterisation techniques are used to analyse the measurement results.

4.1 INTRODUCTION

With the development of many advanced instruments for three-dimensional (3-D) surface topography measurement the applications of 3-D surface metrology is now widespread throughout disciplines such as mechanical and manufacturing engineering, material science, electronic device production, optical surface manufacture, chemical engineering, biological science and bio-engineering. Measured surfaces range from the topography of heavily engineered surfaces[1–8] to single crystals and biological cell surfaces.[9–13] Indeed, the application of 3-D surface metrology has become something of a new challenge in modern science and technology, with its full potential still being largely unrealised.

The vast range of applications of 3-D surface metrology rely on the development of suitably advanced instruments, namely, stylus instrument,[14–16] optical instruments (including focus detection[17–19] and interferometry),[20–22] and scanning probe microscopes (including the scanning tunnelling microscopy (STM),[23–25] and atomic force microscopy (AFM)).[26] A comprehensive review of a full range of topography measurement instruments has been recently carried out by Dong et al.[27] The 3-D stylus instruments which were initiated by Williamson[28] and Peklenik et al.[29] are based on the traditional use of the stylus and are widely used to measure the topography of engineering surfaces. Variants of these instruments have become the most commonly used in manufacturing control, wear, friction and lubrication research. They are accurate, robust and even portable and as a consequence have become the 'gold standard' for instruments used routinely in research and the workshop.

The prevailing advantages of optical instruments over stylus instruments are the non-contacting nature of the data acquisition system as well as the greatly increased measurement speed. In addition, the focus detection instrument which was originally introduced by Minsky[30] and Dupuy[31] has similar characteristics to a conventional stylus instrument in a number of aspects such as measurement range and resolution. Consequently, these instruments have proved suitable for measuring soft materials (soft metals, rubber, coal, paper, magnetic tape and liquid surfaces), surfaces with soft coatings (surface refined sheet metals, magnetic discs), very hard materials (hard metals, ceramics) and elastically deformable workpieces (foils, films). The interferometric instrument has a high accuracy and high resolution. It is best suited for use in precision engineering, where stylus and focus detection instruments are less efficient due to their lower resolution and contact damage problems of the stylus instrument. This type of instrument has been successfully used in measuring diamond machined parts, transparent film surfaces, ball bearings, magnetic media and read/write heads, polypropylene surfaces, super-polished optics, lens moulds, fibre optics, and other optical and precision surfaces.

Since their introduction by Binnig and Rohrer[23] in 1981 (for STM), and Binnig and Quate (for AFM)[26] in 1986, the STM and AFM have become the most widely used instruments for measuring surface topography at the ångström scale. Compared with other 3-D instruments, the STM/AFM is the most applicable to 3-D subnanometer measurement and have become widely accepted both in science and engineering. In biological science the STM/AFM is an excellent tool for visualising DNA molecules and viruses. Biochemical operations are now possible through the ability of the tip to deposit a small amount of material on a given surface. In chemical engineering the STM/AFM allows investigation of structures formed by adsorbates and chemisorbates or chemical surface reactions. In material science crystal structure, growth patterns and crystalline defects can be observed by the STM/AFM. When applied to mechanical engineering the STM/AFM can be used as a micro positionner which enables the tip to be positioned in a specific location on a surface with atomic accuracy. It also serves as a micro indentation machine and a micro-length measurement instrument. Lithography is obtainable by deliberately etching a surface with the STM/AFM which leads on to the development and manufacturing of microcircuits. Currently new areas of application of the STM/AFM are still being exploited in all of the above fields.

In spite of the wide range of applications of 3-D surface metrology, it is still a relatively new technique compared with 2-D surface metrology (profile and area observations) for most researchers and engineers in the relevant fields. To the best of the authors' knowledge, the applications of the technique with the different advanced instruments have not yet been systematically introduced to the wider scientific public. Although research papers and reports are spread throughout the full breadth of technical journals and conference reports, it is difficult for a researcher or a professional metrologist to find specific references which deal with the applications of the different advanced instruments. It is even more difficult for workers new to the field to gain a reasonably comprehensive understanding of the techniques without many hours searching through the scientific literature. Clearly a thorough review of a number of key application areas is needed. This Part is intended to assist in the understanding of measurement techniques and potential applications of 3-D surface metrology through the introduction of a number of examples of measurement applications carried out at the Centre for Metrology at the University of Birmingham. The measurements are carried out using a number of advanced instruments including stylus instruments, focus detection instruments, interferometric instruments and the STM.

The examples introduced are taken from practical research projects carried out at the Centre for Metrology at the University of Birmingham in recent years. They involve measuring gears, engine bores, human hip prostheses, human skin, superconducting thick films and hardness indentations. The instruments used in the applications are (i) stylus type; Talysurf-6 equipped with the lead screw driven stages and DSARG equipped with the linear motor driven stages; (ii) focus detection type, Rodenstock-RM600; (iii) interferometric, WYKO TOPO-3D and (iv) a STM, Burleigh Instructional STM. In order to be able to demonstrate the applications clearly, photographs of *in situ* measurement of surface topography with the different instruments are presented. The mapped surfaces are shown with some visual plots (isometric, contour and grey scale plots) to demonstrate the capability of the instruments to measure proper surfaces and in some cases scanning electron microscope (SEM) micrographs of the surfaces are shown for comparison purposes. Some engineering surfaces are interpreted and characterised through the use of 3-D parameters. In addition, a comprehensive characterisation procedure is illustrated by introducing an example of characterising indentation topography.

4.2 MEASUREMENT OF A GEAR SURFACE WITH THE STYLUS LEAD SCREW DRIVEN INSTRUMENT

Gears are an integral part of many mechanical products. The quality of the gears in engineering applications relies, among other things, on the characteristics of surface topography of the gear teeth. Consequently, the surface topography of the gears is usually one of the important factors controlling the manufacturing process. Most gears are used for loading and movement transmission purposes, and for these types of gears the most suitable instrument for measuring the surface topography is the stylus instrument. The example discussed below shows how a 3-D stylus instrument, based on a modified Talysurf 6, is used to measure such a gear surface.

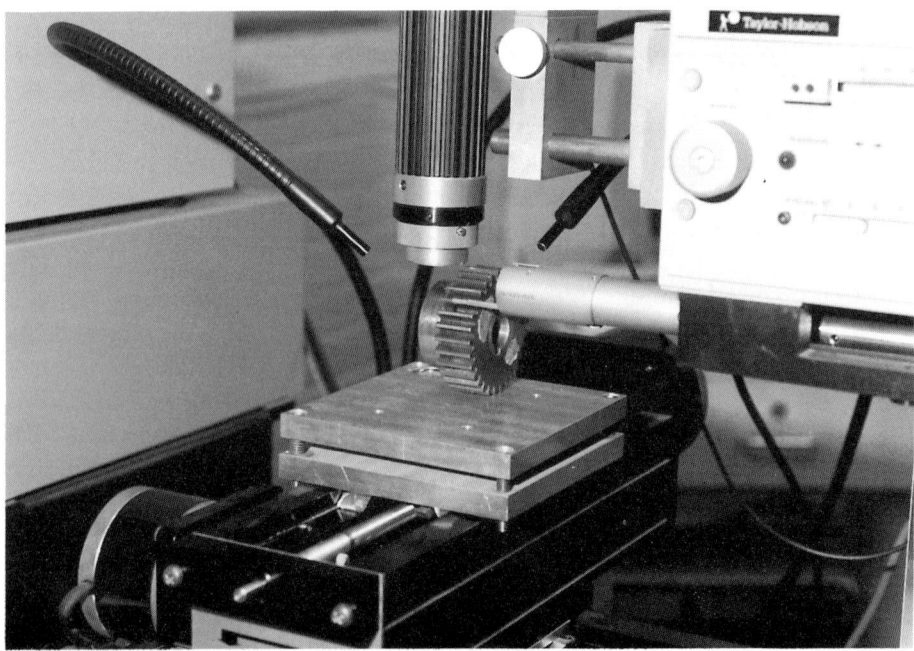

Fig. 4.1 Photograph of a 3-D lead screw driven stylus instrument (modified Talysurf 6) in measuring a gear tooth

4.2.1 Instrument and Topography
Measurement of A Gear Surface

A modified Talysurf 6 (Fig 4.1) is used in conjunction with two orthogonal leadscrew driven translation stages for specimen scanning, and a host computer linked to the original amplifier of the commercial Talysurf 6 analyses the height information. The translation stages are controlled through the host computer, by two stepper motors with a step size of 1.25 µm, a positional accuracy of 0.1 µm and a translation range of 150x150 mm. Backlash problems induced by the leadscrew drive system have been eliminated by moving the work surface on the stages in one direction only for each profile. A static measurement regime is adopted in the system, which avoids the influence of the dynamic characteristics of the stylus on the measurement and eliminates the need to control accurately the velocity of stage movement. The static measurement, however, increases the time required to map a given surface. The specified vertical resolution of the system is defined by the manufacturer in the nanometer scale and the maximum range is 625 µm. The gear is measured by directly placing the specimen on the measurement stage. Fig. 4.1 shows a photograph of the measurement system; in this case a gear tooth surface is being measured.

4.2.2 Characterisation of the Gear Surface

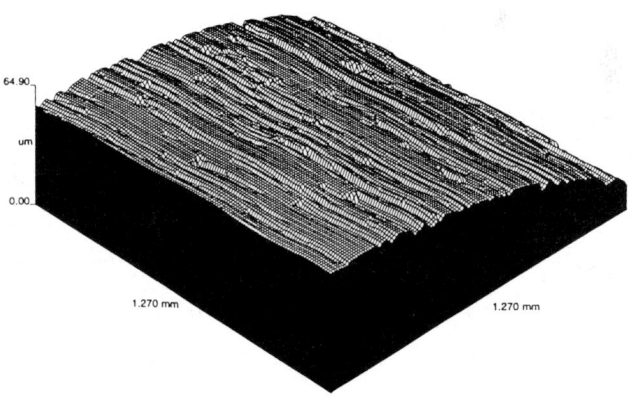

(a) Original topography

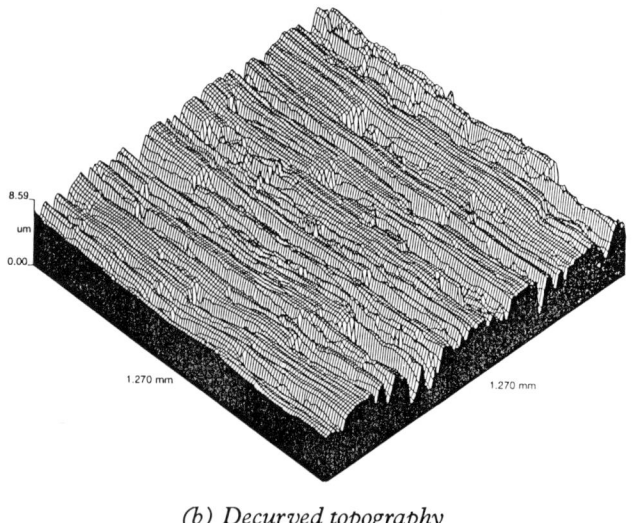

(b) Decurved topography

(c) Grey scale image

Fig. 4.2 Topography of the gear tooth surface measured by a stylus instrument

Fig. 4.2(a) shows the original topography of the gear; wear scars are visible on the plot. In order to see the topographic features in more detail and to characterise the surface roughness, the curvature of the mapped topography has to be removed. The 'decurved' topography and the grey scale image of the gear tooth are shown in Fig. 4.2(b) and 4.2(c) respectively. The figure demonstrates that the wearing process has given rise to a flattening of the machining ridges via sliding/running in wear. The changes of the gear tooth in bearing and the oil retention properties of the surface topography with changes of surface height (bearing area ratio and void volume ratio)[32] are given in Fig. 4.3(a) and Fig. 4.3(b) respectively. The diagram shown in Fig. 4.4 gives an indication of the shape of the height distribution. Since the measured gear is worn, high spikes are almost completely removed, and the consequent topographic height distribution tends towards non-Gaussian distribution with a negative skewness. In addition, characteristics of the surface topography can be further characterised through the use of numerical parameters. Some of the more important parameters with both statistical and functional significance are briefly described below.

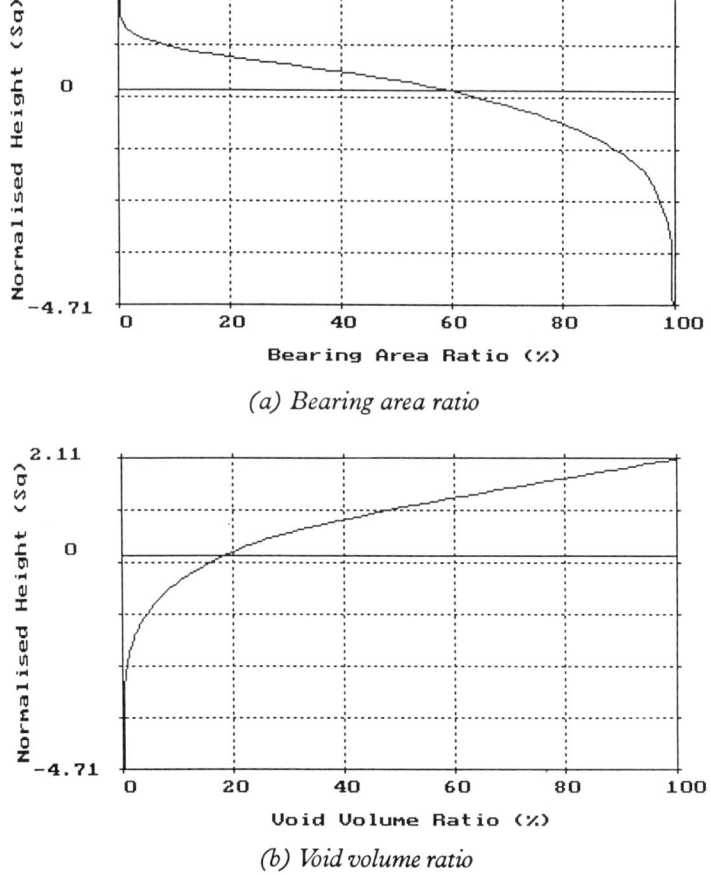

(a) Bearing area ratio

(b) Void volume ratio

Fig. 4.3 Bearing area ratio and void volume ratio of the gear tooth surface

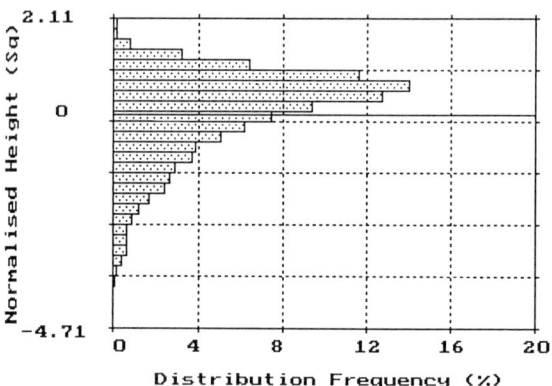

Fig. 4.4 Surface height distribution of the gear tooth surface

S_q – This is the root mean square (RMS) value of the surface departures within the sampling area. It is used to describe the statistical property of the amplitudes of surface topography, and is given by the equation

$$S_q = \sqrt{\frac{1}{MN} \sum_{j=1}^{N} \sum_{i=1}^{M} \eta^2(x_i,y_j)} \qquad (4\text{–}1)$$

where M or N is the number of sampling points in trace direction x or orthogonal trace direction y, and $\eta(x,y)$ is the residual surface obtained by subtracting the reference plane from the original surface.

S_{sk}, S_{ku} – These are the skewness and kurtosis of the surface height distribution. S_{sk} is the measure of the symmetry of the surface deviations about the mean plane. S_{ku} is a measure of the sharpness of the surface height distribution. They are given by the equations

$$S_{sk} = \frac{1}{MNS_q^3} \sum_{j=1}^{N} \sum_{i=1}^{M} \eta^3(x_i,y_j) \; , \quad S_{ku} = \frac{1}{MNS_q^4} \sum_{j=1}^{N} \sum_{i=1}^{M} \eta^4(x_i,y_j) \qquad (4\text{–}2)$$

For a Gaussian surface which has a symmetric surface height distribution about the mean, the skewness and the kurtosis would be 0 and 3 respectively. For other symmetrical height distributions, the skewness remains 0 and kurtosis may be larger or smaller than 3. However, for a significantly asymmetrical shape of surface height distribution, the skewness may be negative if the distribution has a long tail at the lower side of the mean plane or positive if the distribution has a long tail at the upper side of the mean plane. In both cases the kurtosis might be larger than 3. This pair of parameters can be used to indicate some features of engineering surfaces, e.g. outliers and spikes. A comparatively large negative skewness and kurtosis, say $S_{sk}<-1$ and $S_{ku}>5$, may result from surfaces such as ground, honed and plateau honed, which have a few significant outliers (pits, troughs). Conversely, a comparatively large positive skewness, say $S_{sk}>1$, may indicate the presence of small number of spikes on the surface which could quickly wear and become detached during tribological contact.

S_{bi}, S_{ci}, S_{vi} – These are functional parameters and are used to specify bearing, and fluid retention properties of surfaces.[32] S_{bi} is called the surface bearing index and is defined as the ratio of the RMS value over the surface height at 5% bearing area. Usually, the S_{bi} would be larger than 0.7 for a good bearing property. S_{ci} is called core fluid retention index and defined as the ratio of the void volume of unit sampling area of the core zone which is the topography part between 5% and 80% bearing area over the RMS deviation. A large value of S_{ci} indicates a large fluid retention in this zone. S_{vi} is called valley fluid retention index and defined as the ratio of the void volume of unit sampling area of the valley zone which is the topography part between 80% and 100% bearing area over the RMS deviation. A value of 0.15 or larger of S_{vi} indicates a good fluid retention in the valley zone.

The corresponding parameters obtained from the measured gear are given in Table 4.1. Their functional significance will be specified in comparison with the engine bore surface to be introduced below.

Table 4.1 3-D surface roughness parameters of some surfaces

Surfaces	S_a (μm)	S_{sk}	S_{ku}	S_{bi}	S_{ci}	S_{vi}
Gear tooth	1.259	-1.153	4.18	0.872	0.972	0.192
Engine bore	0.765	-3.939	36.26	1.148	0.645	0.199

4.3 MEASUREMENT OF AN ENGINE BORE SURFACE WITH THE STYLUS LINEAR MOTOR DRIVEN INSTRUMENT

The performance characteristics of a combustion engine relies heavily upon surface topography of the engine cylinder bores. As a consequence, the measurement and analysis of engine bore surfaces have long attracted attention in both pure research and manufacturing engineering. The inherent hardness of these surfaces makes them suitable for measurement using a stylus type instrument. The example discussed below uses a linear motor driven stylus instrument, DSARG-3D, in-house constructed at the University of Birmingham, to measure an engine bore surface.

4.3.1 Instrument and Topography Measurement of An Engine Bore Surface

DSARG-3D is based upon the same principles as the Talysurf 6 3D system in terms of the stylus. However, its translation stages are different from that of Talysurf 6. Instead of using a lead screw driven mechanism as in Talysurf 6, the DSARG-3D uses a linear motor driven mechanism which eliminates backlash and any possible movement errors of the stages caused by the lead screw. The stages in combination with the stage control system are able to achieve highest spatial resolution of 0.78 µm, and horizontal measurement range 100x100 mm. The highest vertical resolution and maximum measurement range of the gauging system are 4.6 nm and 1mm respectively. Both static and dynamic (on-the-fly) measurements are possible using this instrument. Due to its wide measurement range and high spatial resolution, the linear motor equipped stylus instrument is suitable for measuring a wide range of engineering surfaces. Fig. 4.5 shows a photograph of the system in use.

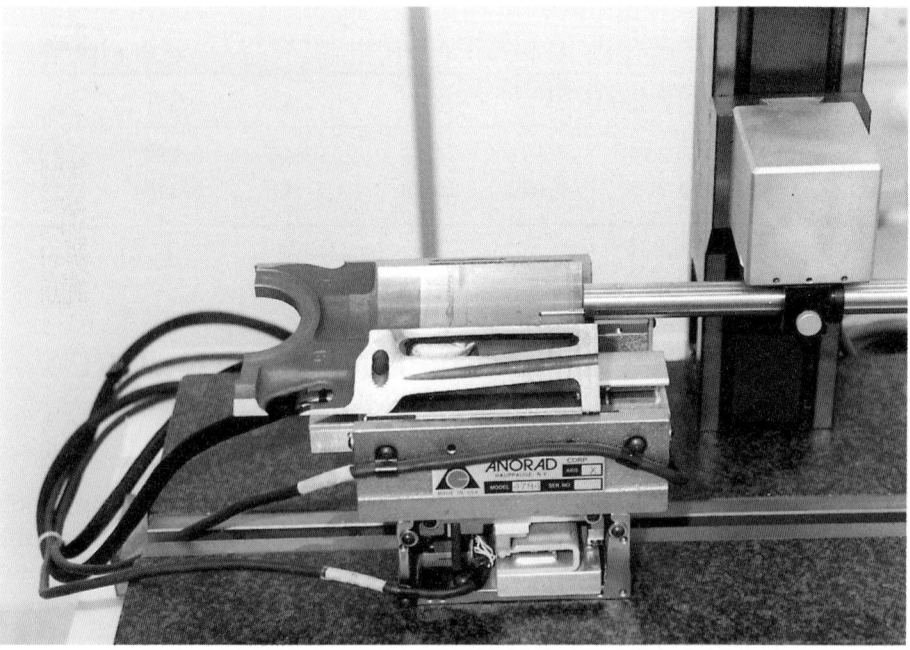

Fig. 4.5 Photograph of a 3-D linear motor driven stylus instrument (DSAGE-3D) in measuring an engine bore

4.3.2 Characterisation of the Engine Bore Surface

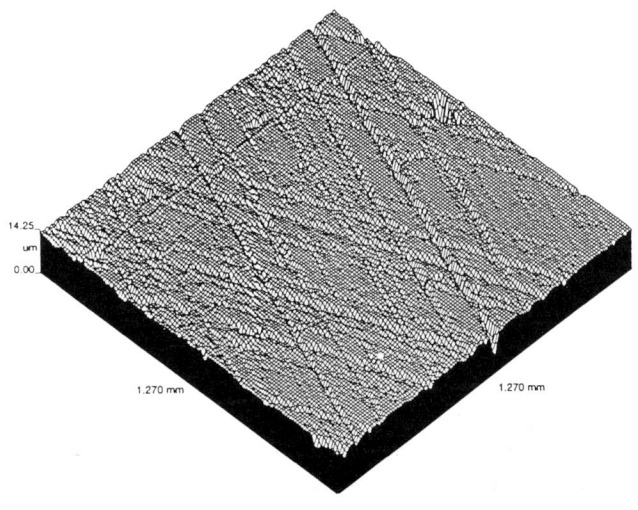

(a) Decurved topography

(b) Grey scale image

Fig. 4.6 Topography of the engine bore measured by a stylus instrument

In a similar manner to the gear surfaces, the cylindricity of the engine bore surface has to be removed from the collected data before the topographical features can be readily observed. Fig. 4.6(a) shows the 'decurved' surface topography of the engine bore surface. Typical features of engine bore topography are clearly visible. Analysis of the topography shows that it possesses a relatively smooth upper surface zone with deep penetrating machining grooves in the lower surface levels; these are characteristics of the honing and/or plateau honing process used in the manufacture of the engine bore. The relatively flattened nature of the surface zones is readily observable from the skewed graph of the height distribution of the surface asperities, shown in Fig. 4.7. The highly grooved nature is also identifiable from the grey scale image (Fig. 4.6(b)) of the surface. Functionally, the engine bore surface is designed to have good bearing and oil retention properties. The flattened top zones and the highly grooved machining marks match the functional requirements in that the contact load is supported by the flattened areas while the oil can flow along the network of deep grooves providing a suitable supply for hydrodynamic lubrication. The bearing area ratio and the void volume ratio are shown in Fig. 4.8(a) and 4.8(b) respectively. From a parametric point of view, a high quality engine bore surface should have a large bearing index (S_{bi}>0.7), a large valley fluid retention index (S_{vi}>0.15) and a suitable core fluid retention index (S_{bi}>0.4). As compared with the parameters of the gear tooth, the parameters of the engine bore surface shown in Table 4.1 indicate better functional properties in terms of the bearing property. They have similar fluid retention property in the valley zone, but the gear tooth possess a better core fluid retention index.

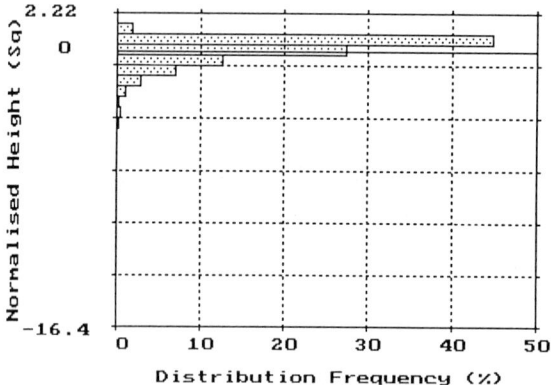

Fig. 4.7 Surface height distribution of the engine bore surface

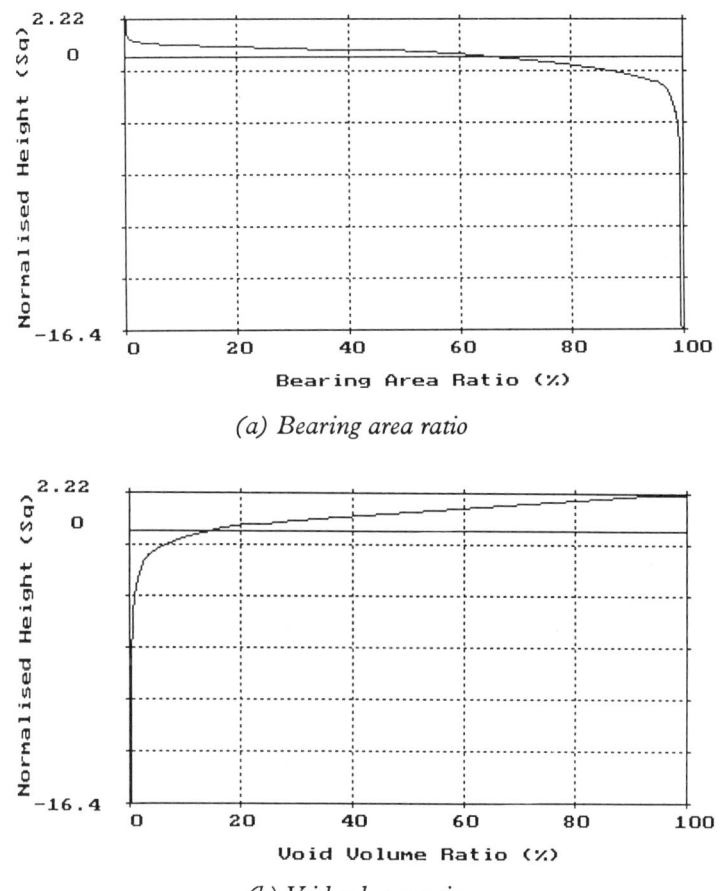

(a) Bearing area ratio

(b) Void volume ratio

Fig. 4.8 Bearing area ratio and void volume ratio of the engine bore

4.4 MEASUREMENT OF THICK FILM SUPERCONDUCTORS WITH THE FOCUS DETECTION INSTRUMENT

In recent years optical profilometry has become a promising technique in 3-D surface measurement. Fast measurement speed and non-contact measurement are the main advantages of optical instruments over mechanical stylus instruments. In this example a focus detection instrument (Rodenstock-RM600) is used for the measurement of a superconducting $YBa_2Cu_3O_{7-x}$ (YBCO) material which would be seriously damaged by contact measurement with a stylus.

4.4.1 Instrumentation

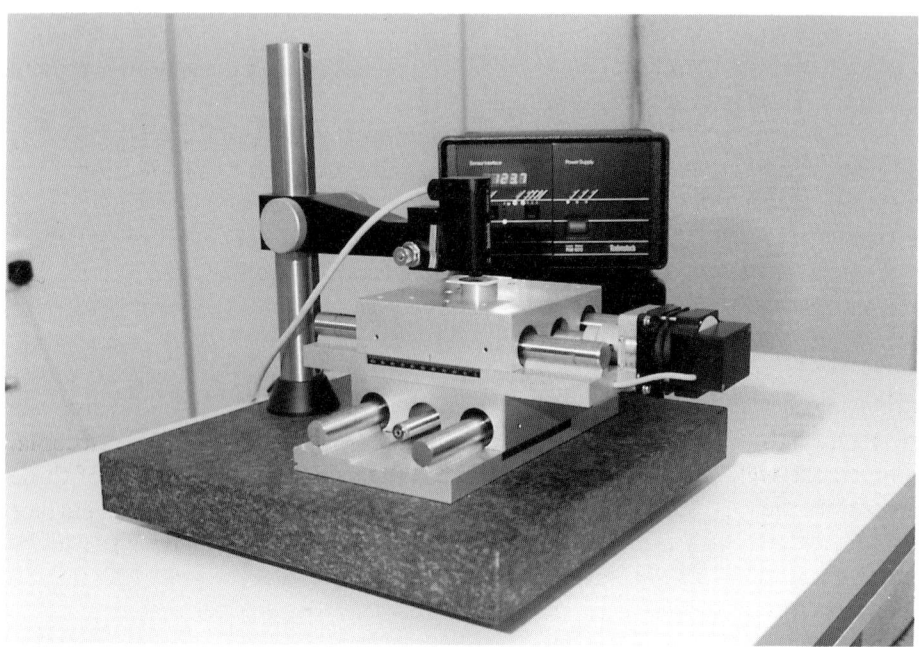

Fig. 4.9 A 3-D focus detection instrument (Rodenstock 600) used in measuring superconducting material

The Rodenstock RM600 is an optical focus detection instrument. As is seen from Fig. 4.9, the system consists of a static measurement head and two precision lead screw driven orthogonal translation stages for realising 3-D measurement. The horizontal driving mechanism results in a large scanning range of 300x300 mm. The highest horizontal resolution of the instrument is restricted by the combination of the size of the light spot and the resolution of the translation stages. A significant advantage of the instrument is that the stand off distance (a distance between the optical head and the surface measured) can be up to 15 mm which makes some applications more convenient. The highest scanning rate is up to 600 Hz with a dynamic vertical resolution 0.05 μm, and a maximum vertical measurement range of 600 mm. The system can carry out raster scans in either the x or y axis using an 'on the fly' data sampling regime. Fig. 4.9 shows a photograph of the system measuring a thick film superconductor.

4.4.2 Fabrication of the Thick Film Superconductors

It is a well established fact that surface resistance is dependant upon surface roughness; a rough surface has a high surface resistance and a smooth surface has a relatively low surface resistance. [35] When considering the surface resistance of superconducting thick films it is essential that the topography of the thick films should be quantified.

With advances in the fabrication and refinement of $YBa_2Cu_3O_{7-x}$ (YBCO) superconductors, attention has focused on the methods of production. One of these methods is the screen printing technique.[36] Screen printing is a well proven technology and is at present widely used in the production of hybrid electronic devices with a large variety of material combinations. Fabrication of superconducting thick film tracks by this technique is particularly appropriate. In this case YBCO powders were prepared by high temperature solid state synthesis of Y_2O_3, CuO, and $BaCO_3$ powders and the synthesised material was ball milled and prepared as an ink. The screen printing was performed through a 325 mesh on to substrates of yttrium-stabilised zirconia. The tracks were then sintered at 975°C. The relatively soft nature of these materials means that stylus type measurements would damage the surface there; hence a non-contact method is the only real option for topography measurement. Thus the instrument employed here was Rodenstock RM600 – a non-contact optical instrument.

4.4.3 Topography of the Thick Film Superconductors

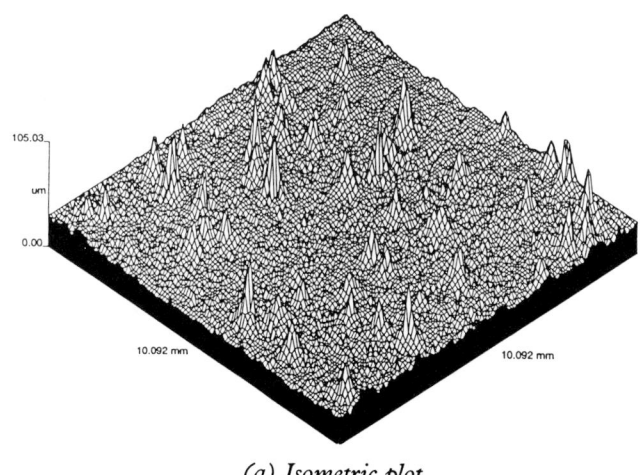

(a) Isometric plot

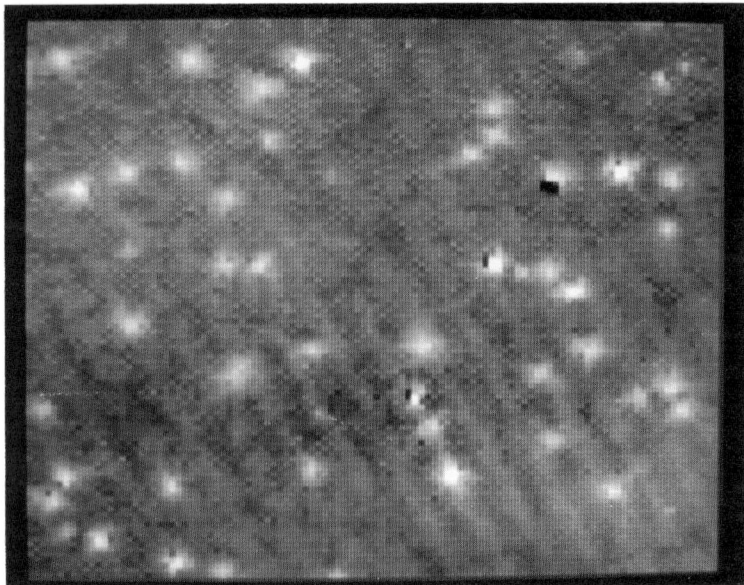

(b) Grey scale image

Fig. 4.10 Topography of the high surface resistance specimen measured by a focus detection instrument

Fig. 4.10 shows a 3-D surface map and an intensity plot of the surface features of the high surface resistance specimen. Clearly visible are the previously reported[36] preferential crystal growth sites where the material has formed peaks of various elevation. In between these peaks, the regular texture imparted from the mesh can also be seen. It should be noted that the peak heights on the surface of this particular specimen are of the order of 100 μm. When the surface of the specimen possessing the lower surface resistance was measured, Fig. 4.11, analysis showed that the general level of the preferential growth sites are considerably lower, being of the order of 50 μm. Also clearly evident on this surface are low elevation pathways across the surface (dark paths on Fig. 4.11); these are known as rosette boundaries[36] and surround areas of preferred crystal orientation. These pathways and growth rosettes are typical features of these types of thick film superconducting surfaces. Optical and SEM micrographs are shown in Fig. 4.12(a) and (b) for comparison purposes and they indicate that the majority of the features are recorded by the focus detection system although the small re-entrant features associated with the preferential crystal growth sites, visible using an SEM in figure 4.12(c), are not distinguishable using 3D topographical analysis. Further analysis of these paths could be accomplished by zooming, truncation and clipping of some of the larger scale features.

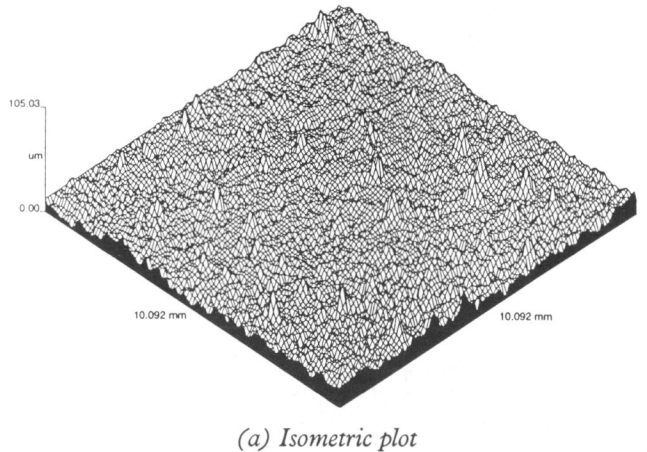

(a) Isometric plot

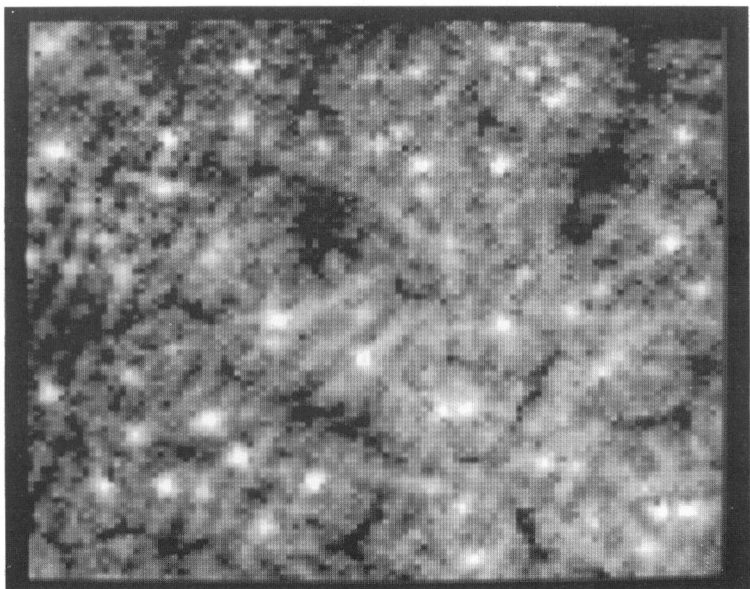

(b) Grey scale image

Fig. 4.11 Topography of the low surface resistance specimen measured by a focus detection instrument

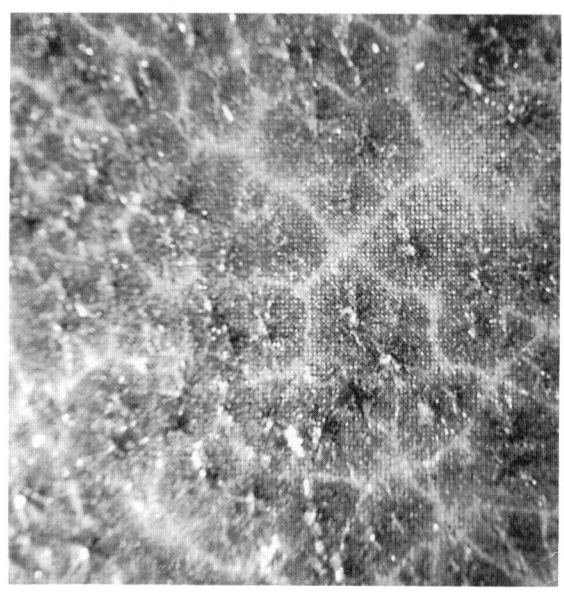

(a) Optical micrograph of the high surface resistance specimen

(b) SEM image of the high surface resistance specimen

(c) SEM image of the low surface resistance specimen

Fig. 4.12 Topography of the superconducting material measured by a SEM and an optical microscope

The above analysis gives a distinct indication of the topography differences encountered between high and low contact resistance films and indicates clearly the potential of the focus detection technique for relatively quick and damage-free examination of the topographical surface features.

4.5 MEASUREMENT OF HUMAN SKIN WITH A FOCUS DETECTION INSTRUMENT

In addition to the wide range of applications in engineering, 3-D surface metrology is also applicable in a number of other non-engineering fields. These include biological surfaces such as skin, membranes, plant leaf surfaces and cellular structures. For example, when considering skin cancer, accurate measurement of the tumour size and the consequent rate of growth of the tumour are closely related to the prognosis of the patient.[33] Commercially, manufacturers of cosmetic skin creams and lotions make various claims about their product's ability to enhance the skin's aesthetic topographical appearance. Both of the above examples highlight areas of application for quantitative application of 3D topography measurement.

Clearly the relatively soft skin surfaces considered above are suitable only for non-contact measurement. To measure skin requires the use of replicas of 'soft' silicone rubber. The example discussed here shows 3D maps of the skin at the end of the finger tip measured from a rubberiod replica using a focus detection instrument (Rodenstock RM600).

4.5.1 Replication of Surfaces

In general, when measuring surface topography, direct measurement of the surface can usually be made; however in certain circumstances where the specimen is too 'soft' or because of the difficulty in assembling the part on to the working area of the instrument a surface replication technique must be used. This usually involves allowing a hard setting liquid to flow into and take up the shape of the surface. When the replica is hardened it is removed and measured giving a 'mirror image' replication of the specimen topography.

When measurements are made of human skin, replication techniques are required due to both the difficulty of getting the skin, as a specimen, on to the instrument and the fact that skin is a 'soft' material. Clearly, the replication material must be clinically safe. In the example used here, Exaflex, a vinyl silicone rubber material widely used in dentistry, was used. Measurements were made using the non-contact focus detection technique employing the Rodenstock RM600 instrument.

4.5.2 Topography of the Skin Surface

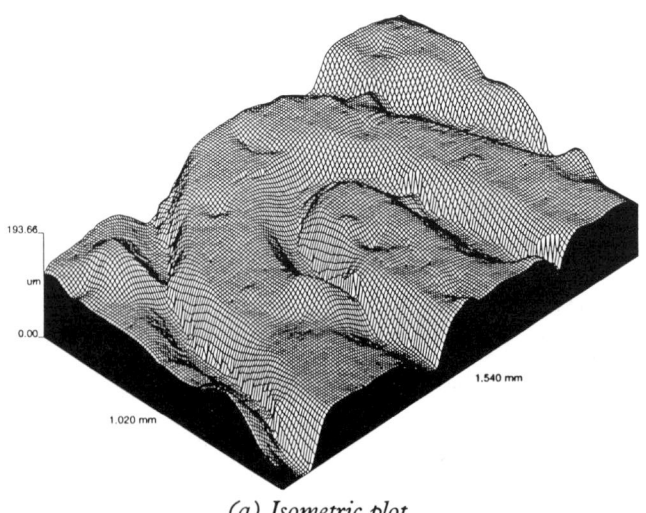

(a) Isometric plot

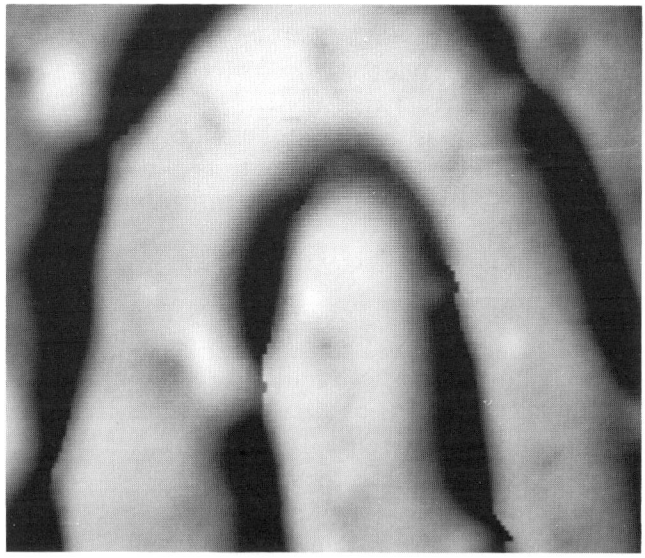

(b) Grey scale image

Fig. 4.13 Topography of a fingerprint measured by a focus detection instrument

An isometric plot and a grey scale image of a fingerprint are shown in Fig. 4.13(a) and 4.13(b) respectively. Not only can the pattern of the fingerprint be observed, but also the detail dimension in all directions can be calculated from the digitised topography. For example, the volume of a protruding tumour could be calculated along with the volume of the troughs and crevices within the surface. Moreover the digitised topography can be further processed by zooming, inversion, truncation, and clipping. The maps clearly show the potential of this kind of measurement in the biological field, especially where quantitative topography measurement is required. A number of projects are currently being carried out in this area.[34]

4.6 MEASUREMENT OF THE TOPOGRAPHY OF HIP PROSTHESES USING PHASE SHIFTING INTERFEROMETER

The replacement of the human hip joint with artificial joints is now routinely undertaken. The consensus view, however, is that unless the ball area of the joint is exceptionally smooth, excessive wear takes place in the joint area, necessitating an expensive replacement operation. The production of the surgical steel artificial joint is very expensive with the ball zone requiring a high degree of polishing. The degree of polishing is such that an accurate measurement of the surface topography is best made using an interferometric measuring system, in this case the WYKO TOPO 3D. The following example details the attempt to correlate surface topography measurements of two hip joint ball surfaces with their service performance. An expensive yet successful implant and a relatively less expensive implant with a poorer service record are considered.

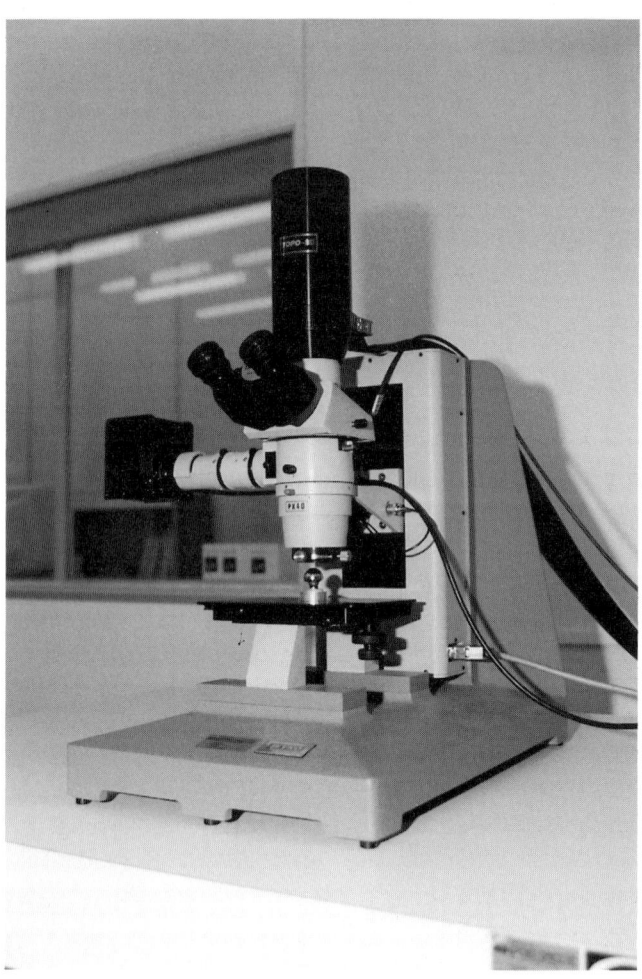

Fig. 4.14 Photograph of a 3-D interferometric instrument (WYKO TOPO-3D) in measuring the hip prostheses

4.6.1 Instrument and Topography Measurement of Hip Prosthesis Surfaces

WYKO TOPO-3D is a phase shifting[37] interferometric instrument. Its vertical resolution is of the order of 0.1 nm, making it an ideal instrument for the accurate measurement of super-polished and other precision surfaces. In principle, the instrument carries out a parallel mapping of surface topography by detecting the phase shift of the interference pattern created between the specimen surface and a reference mirror with a high density imaging array. No scan is needed in the measurement. The horizontal resolution of the instrument is dependent on the magnification of the objective used and the density of the imaging array. With a 40x objective, the horizontal resolution is 1 μm for TOPO-3D. Similarly, the horizontal measurement range is fixed as long as the objective and the imaging array are fixed. A maximum area of 1 mm² is available for the TOPO-3D. A considerable advantage of this instrument over the stylus and the focus detection instruments is its areal mapping speed. The measurement speed is of the order of seconds while for both the focus detection and especially the stylus mapping techniques the measurement time is in the range of tens of minutes. Fig. 4.14 shows a photograph of the WYKO TOPO-3D system being used to measure the ball surface of one of the hip joints.

4.6.2 Characterisation of the Hip Joint Surfaces

Fig. 4.15 and 4.16 show typical maps of successful and poor hip joint surfaces respectively. A number of measurements were carried out and the average surface roughness values were found to be remarkably similar. Closer examination of the surface maps, however, revealed that the surfaces were quite different in nature. The successful surface was exceptionally smooth with relatively large pits in the surface giving the height distribution a skewed character. The smooth zones were measured with the pits deleted from the calculations and found to be of the order of 5 nm. The poor surface displayed evidence of scratches from the polishing operation and also high spots or peaks on the surface. The investigation reveals that the excessive wear that occurs in the poorer quality joint results from inferior polishing operations; in practice it is considered that the high spots may become detached acting as abrasive particles and that the scratch edges could also act as point where excessive wear will take place.

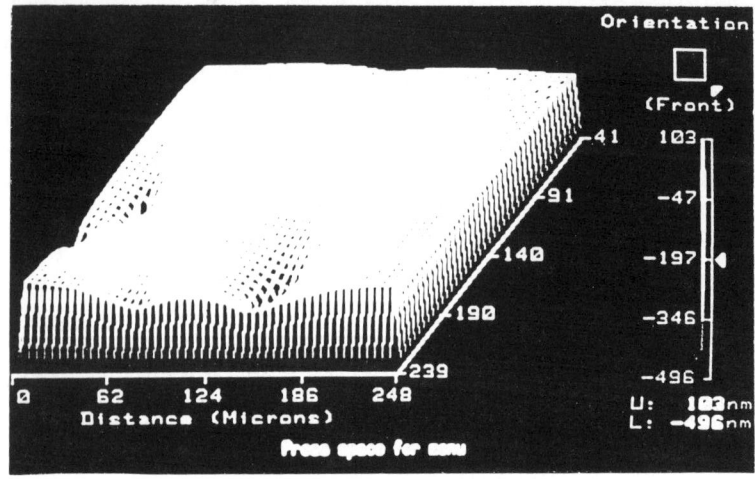

(a) Isometric plot

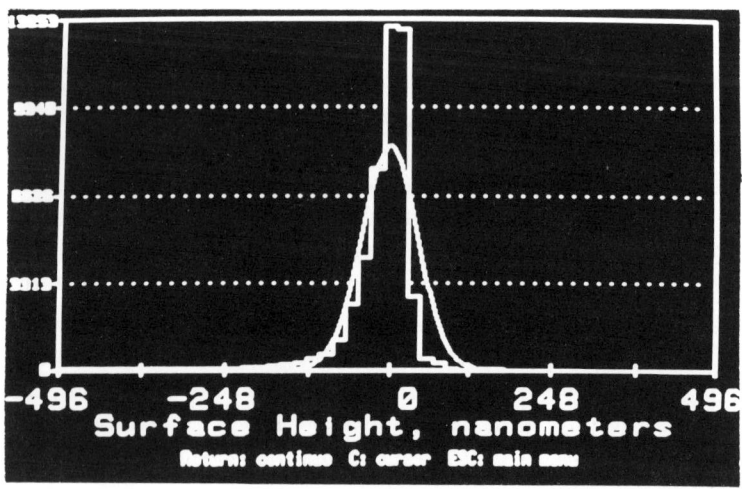

(b) Height distribution

Fig. 4.15 *Topography of a successful hip joint measured by an interferometric instrument*

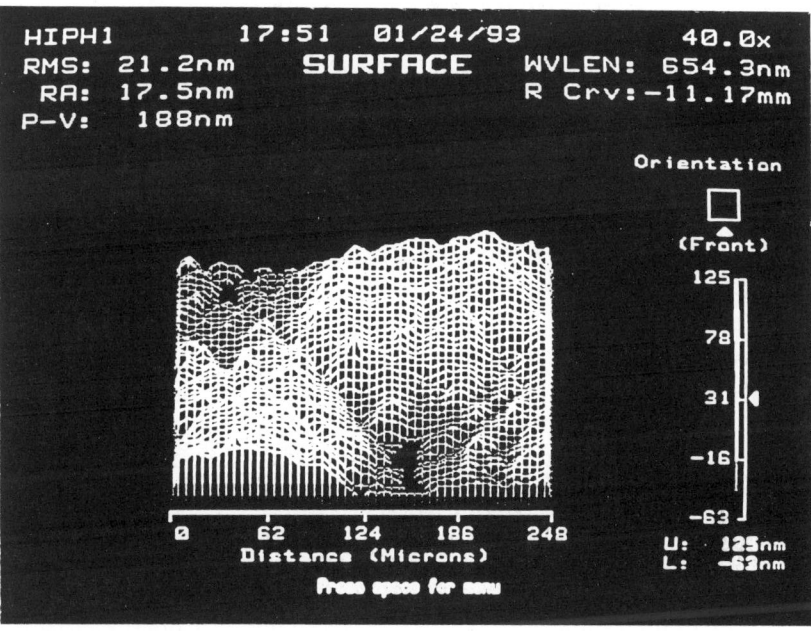

Fig. 4.16 Topography of a poor hip joint measured by an interferometric instrument

4.7 MEASUREMENT OF POLISHED BRASS SURFACE USING A SCANNING TUNNELLING MICROSCOPE

Many manufactured surfaces are polished as a final finishing operation and the resulting surface shows very fine surface features. Clearly the polishing operation must leave 'machining' marks on the surface; in practice, however, these are usually not resolved using conventional 3D surface measurement techniques. The main problem is that although the vertical resolution of some stylus and, more routinely, optical instruments is in the nano-metre range, the lateral spacing of both stylus and optical instruments is too large to measure the scratches left by the polishing operation to be clearly resolved. This makes quantitative surface analysis of the polishing scratches produced by very fine polishing mediums almost impossible. The solution to this problem is through the use scanning probe microscopy. This technique allows lateral resolution down to the sub-nanometre range and hence provides a ready tool for quantitative analysis of this type of surface. The resulting surface topography of a 70:30 brass material polished using a 1 µm, diamond polishing medium is discussed below.

4.7.1 Instrument and Surface Measurement
of a Polished Brass Surface

Ultimate vertical and horizontal resolution at the sub-ångström and ångström level is attained through the use of scanning tunnelling microscopes (STM) and atomic force microscopes(AFM). The scanning tunnelling microscope was pioneered by Binnig.[23] In principle a conducting probe tip of nominally one atom diameter is driven to within nanometers of the specimen surface. A bias voltage of 2 mV-2 V is then applied across the gap and electrons tunnel across the gap. The monitored current is of the order of pA – nA. This current increases exponentially as the gap is decreased and for a 1 ångström gap change the tunnelling current changes by an order of magnitude. This sensitivity allows vertical resolutions of 0.01 ångströms. The scanning mode is usually based around a constant current feedback regime and raster scanning. The x, y, and z motions are provided by a tripod configuration of pizeo-electric elements with recent instrument employing a pizeo tube set-up for added speed and stability[24] this system allows lateral resolutions of 1 ångström. A maximum of 5 µm is usually claimed for the vertical range and a lateral range of 100x100 µm.

One of the main limitations of the STM is that it is only possible to measure conducting surfaces and this proved to be one of the driving forces behind the development of the atomic force microscope by Binnig and Quate. [26] In this case an ultra fine diamond tip is scanned across the specimen surface recording the inter atomic forces between the tip and the atoms of the sample. The tip actually touches the sample and the mode of operation is much like that of a conventional stylus instrument. The tip force is tiny, about 10^{-6} – 10^{-9} N, and at such low forces the tip can trace over atoms without damaging the surface. The tip can be made from a fractured diamond fragment and is attached to a cantilever system. The cantilever is small and has high resonant frequencies, a typical cantilever of silicon oxide having a resonant frequency of 100 KHz. The deflection of the cantilever can be measured using an electron tunnelling microscope (STM), an interferometer or as in a number of commercial instruments, by deflection of a laser beam reflected off a mirror mounted on the back of the cantilever. All that is required is an electrical signal that varies rapidly with deflection. The signal is sent to the same electronic system as used for the STM. Specifically, a feedback circuit controls the voltage applied to the z piezo element so that the signal is held constant as the tip is scanned across the surface. The x, y scanning mode is the same as that employed for the STM i.e. piezo tripod or tube. The AFM is used mainly to measure non-conducting organic and biological materials. A photograph of a STM is shown in Fig.4.17.

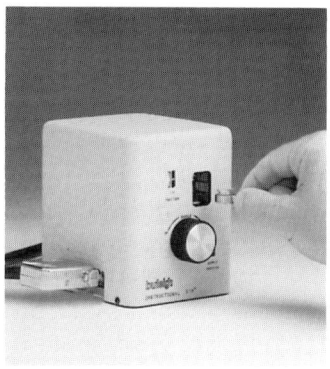

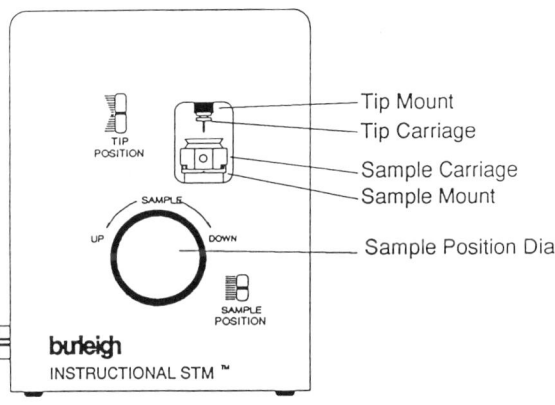

Fig. 4.17 Photograph of a STM (Burleigh instrument)

4.7.2 Topography of the Polished Brass Surface

For measurement of the brass surface a cleaved graphite tip was used over a scanning area of 3 μm x 3 μm and 7 μm x 7 μm. Fig. 4.18(a) shows a grey scale contour map of the surface showing a relatively large dark polishing scratch across the diagonal of the surface. An axonometric view of the same feature is shown in figure 4.18(b); in this case the scale and the width of the scratch is clearly visible against the level of the surrounding polishing scratch marks. When the area map has been collected the data can be manipulated by a host microcomputer. Figure 18(c) shows a single trace extracted from the areal data; the orientation of the trace is such that it crosses the deep scratch orthogonally. From the single trace it can be seen that the scratch is of the order of 1 μm in width and 0.1 μm in depth. Additionally, it can be seen that lateral topographic features of the order of nanometers can be resolved within the scratch. Such one-off deep scratches are not untypical of this type of polished surface and may be reduced through the use of a finer, 0.25 μm Al_2O_3 polishing medium.

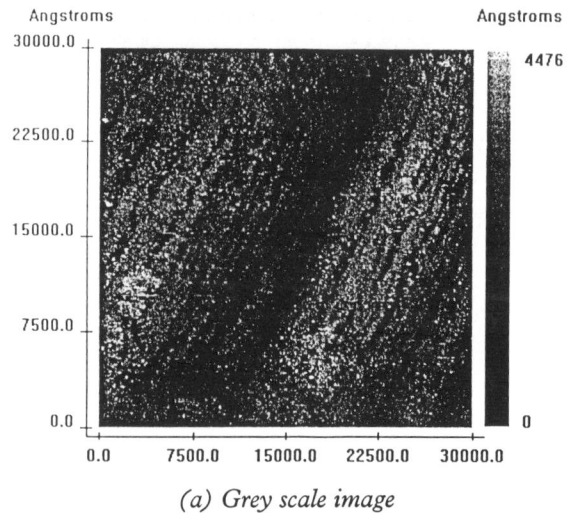

(a) Grey scale image

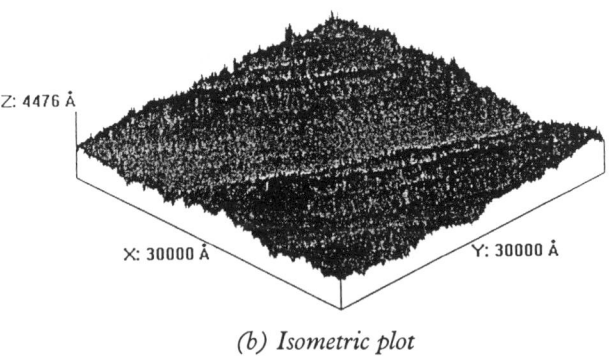

(b) Isometric plot

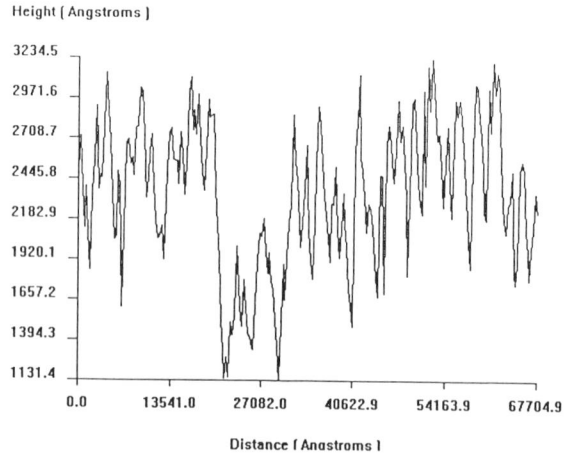

(c) Single trace

Fig. 4.18 Topography of a brass surface measured by a STM

4.8 CHARACTERISATION OF SURFACE TOPOGRAPHY OF INDENTATIONS

Above we have introduced a variety of 3-D instruments and their applications in measuring different kinds of surfaces, and some simple characterisation techniques have been illustrated. Clearly, to measure surface topography is not the final objective of 3-D surface metrology. To characterise the surface i.e. to understand what the user has obtained is the final goal. Therefore in the last section of this Part, we will outline a more detailed example of the use of true 3D topography analysis. The example discussed below is the characterisation of the surface topography of hardness indentations and it illustrates the methodology of the characterisation techniques comprehensively. The intention is to introduce a general procedure for all applications which require sophisticated visual and numerical characterisations of the surface topography, e.g. quality inspection of machined parts, wear testing and other tribological systems.

A general and formalised method for testing the hardness of materials is to indent a surface and then to inspect the indentation topography. In fact, the characterisation of indentation topography is of major importance both in the understanding of the mechanism of the indentation process, and revealing additional information regarding the mechanical behaviour of the test material. The study of indentation topography involves analysing the surface disturbance, including the well-known phenomena of material 'pile-up' and 'sink-in'. Traditionally, this was carried by means of the scanning electron microscope (SEM), multiple beam interferometry, 2-D profilometry[7] qualitatively, and by an optical eyepiece and adjustable crosshairs,[8] measuring the indentation diameter and inferring the contact area, quantitatively. Use of all of the traditional techniques resulted only in partial characterisation. Recently, Sullivan and Blunt[7,8] have adopted the advanced characterisation techniques in 3-D surface metrology and made a comprehensive examination of the characteristics of the indentation topography.

In order to specify the visual and numerical characterisation techniques, indentations were made by carrying out hardness tests using a Lietz Miniload Vickers type microhardness tester with a load of 5 N and a dwell period of 15 s. The shape of the diamond indentor of the microhardness tester is a square based pyramid having an included angle of 136°, and the specimen material was mild steel. The indentation was measured using the modified 3D Talysurf-6 based system with the sampling interval of 10x10 µm and the number of sampling points 80x80.

4.8.1 Visual Characterisation

Visual characterisation of a surface means that the geometrical characteristics of the surface topography are qualitatively assessed through visual perception; it is an intuitive process based upon the experience of the worker, aided by a number of graphical plots. Some of the graphical plots have been introduced separately in preceding sections; in the following these techniques are now summarised via the characterisation of indentation topography.

- The general property of the indentation can be seen from the isometric plot shown in Fig. 4.19. The basic properties of the indentation, namely the pile-up, sink-in and the shape of the diamond of the microhardness tester, are easily observed from the isometric plots at different projection angles.

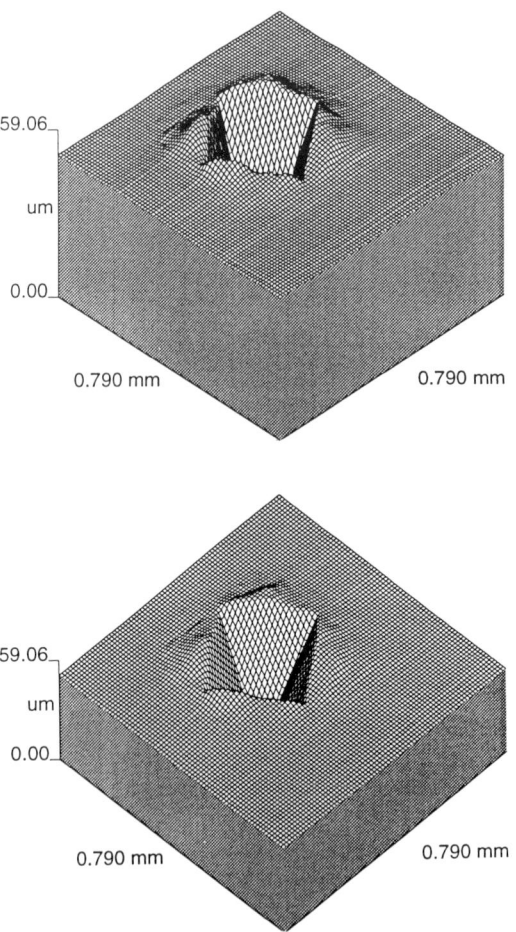

Fig. 4.19 Isometric plot of the indentation

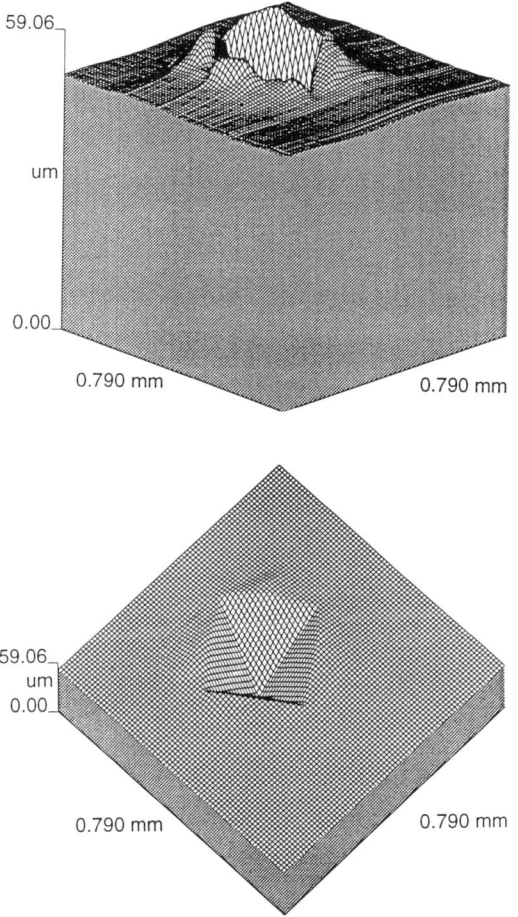

Fig. 4.19 Isometric plot of the indentation (continued)

- The intensity plot, i.e. grey scale image, (Fig. 4.20) presents a near-photographic image of the indentation. The grey scale image is similar to that observed using microscopy; however, it has an inferior vertical resolution and in this figure the pile-up is hardly resolvable. However, the form of the pyramidal indentation of the diamond of the micro-hardness tester is very clear and would allow relatively accurate horizontal dimensional measurement to be carried out.

Fig. 4.20 Intensity plot of the indentation

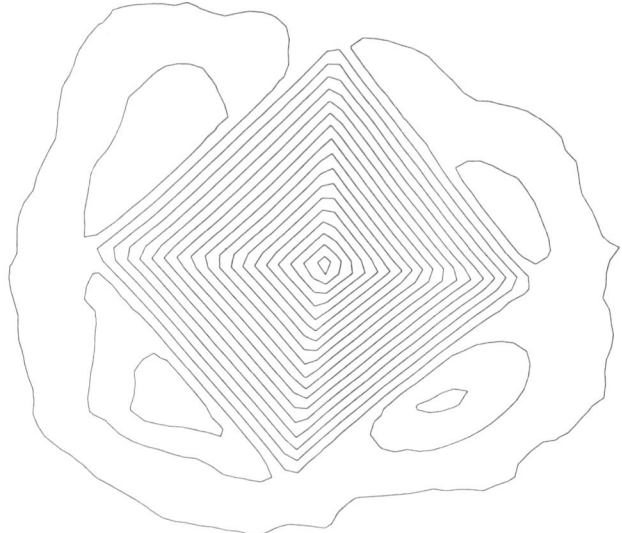

Fig. 4.21 Contour plot of the indentation

- The significant advantage of the contour plot of the indentation is that it provides information of a directional nature relating to the surface topography. The contour plot of the indentation shown in Fig. 4.21 shows the asymmetry of the indentation clearly. It suggests that the disturbance produced around the edge of the indentation is not regular. Asymmetric disturbance is thought to be related to crystalline plane direction and could assist in realising crystal orientation.

- In order to see the topographic details of the part of the indentation below the general level of the surface, an inverted isometric plot is shown in Fig. 4.22. It exhibits clearly the plastic deformation of the material. The pile-up is seen from the front view of the indentation. Just below the level of the surface there is evidence of small amounts of sink-in before the shape of the indentation follows exactly the shape of the diamond of the microhardness tester.

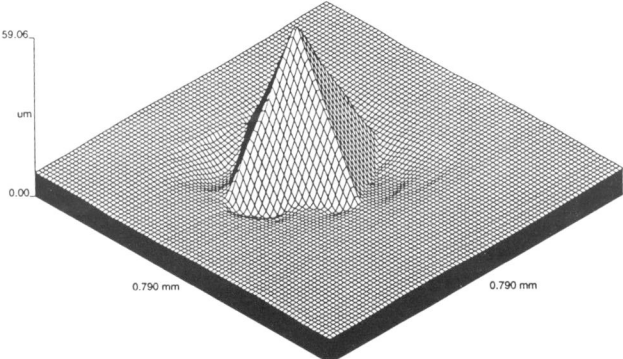

Fig. 4.22 Inverted isometric plot of the indentation

- To analyse the pile-up behaviour in more detail, the indentation may be truncated at the original surface level leaving anything above that level clearly visible, (Fig. 4.23). The figure clearly shows the two distinct zones of pile-up located at opposite sides of the indentation extending across the face of the diamond impression.

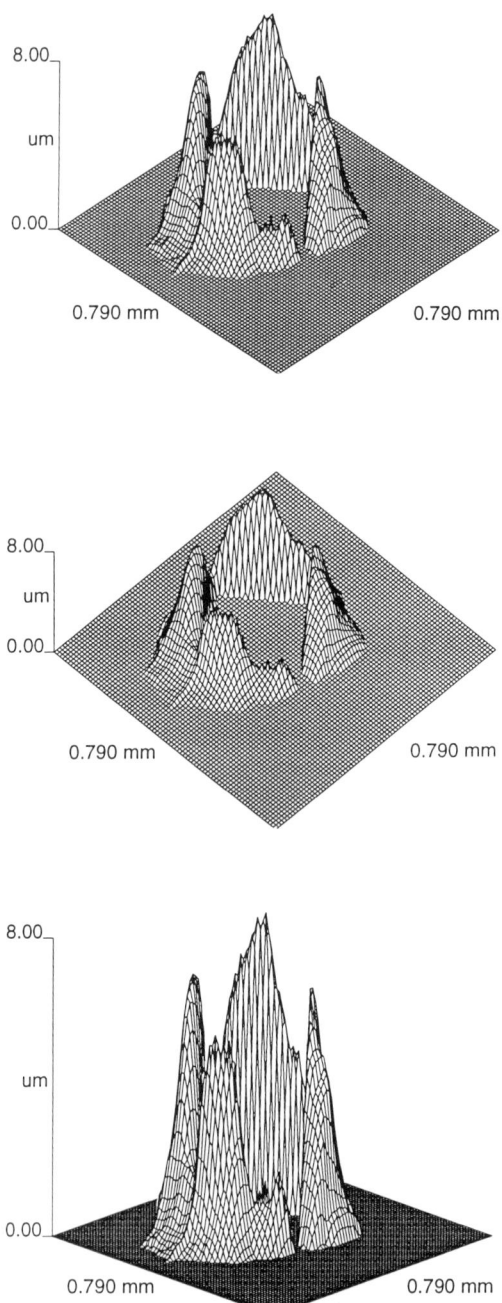

Fig. 4.23 Displaying the pile-up of the indention by combining inversion and truncation

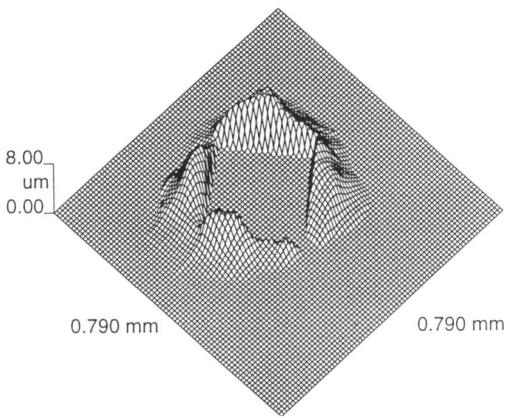

Fig. 4.23 Displaying the pile-up of the indention by combining inversion and truncation (continued)

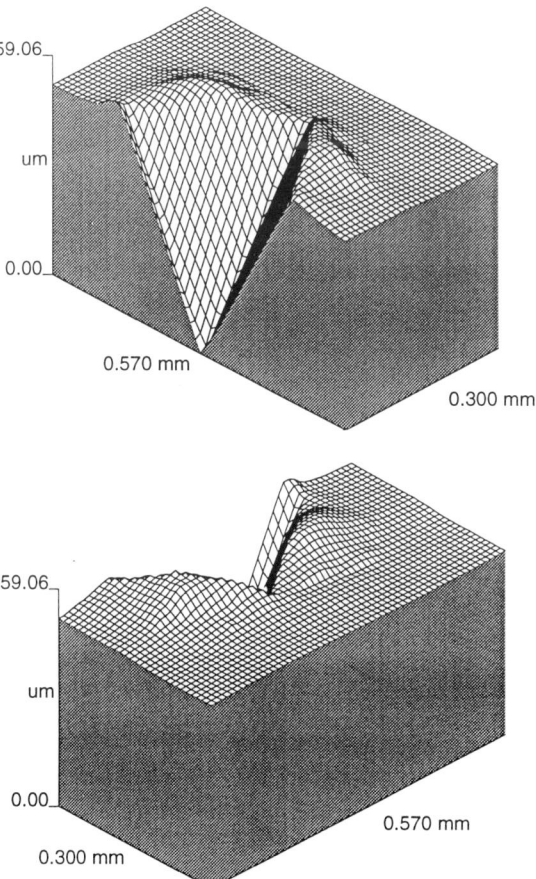

Fig. 4.24 Displaying the zoomed area of the indention

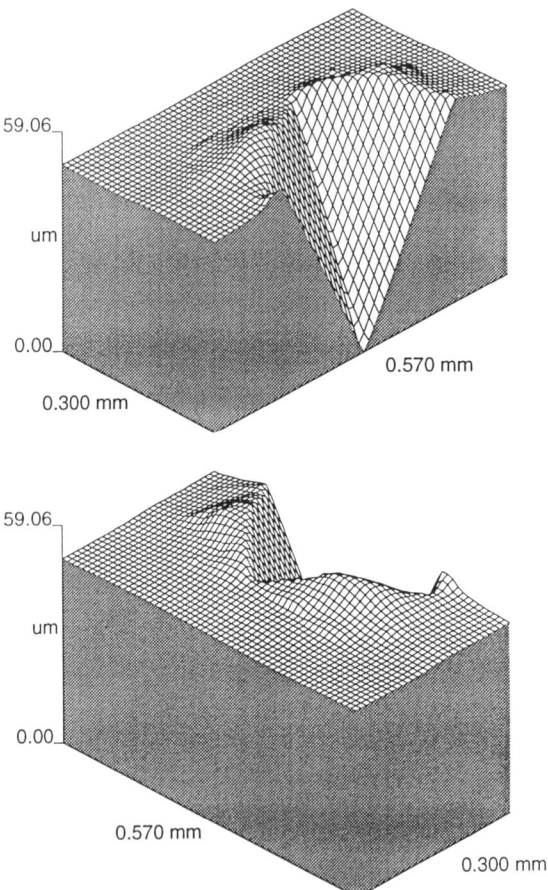

Fig. 4.24 Displaying the zoomed area of the indention (continued)

- For further viewing of the topography of the indentation, the indentation may be zoomed in such a way that only a small area of the indent is displayed. Fig. 4.24 shows a zoomed indentation where the gradual change of the sink-in is seen clearly. This type of analysis is particularly useful for detecting flaws in the diamond indentor.

4.8.2 Numerical Characterisation

The above visual characterisation technique can only give a qualitative assessment of the indentation. Traditional quantitative characterisation techniques, e.g. the use of an optical eyepiece and adjustable cross-hairs measuring the indentation diameter and inferring the contact area, can not provide accurate numerical calculation. With 3D surface metrology, however, the geometrical quantities of the indentation, e.g. the depth of the indentation, the spread area and the volume of the pile-up can be obtained accurately. From the digitised indentation topography, first of all, it is easy to obtain any of the numerical parameters adopted in the above sections and proposed elsewhere.[32,34] Secondly, further analysis for specific topographical features can be carried out.

When considering indentation characterisation specific analysis is preferable to the roughness parameters, so the total height (from the top of the pile-up to the bottom) of the indentation, which is 59.1 μm in this case, is calculated. Then the bearing area ratio, extrusion volume (material volume) ratio and the void volume ratio[32] are plotted against surface height as shown in Fig. 4.25. As is seen from the bearing area ratio curve (Fig. 4.25(a)), there are three distinctive stages. The first stage is the slow increase of the bearing area as the truncation level goes down. This corresponds to the change of the bearing area of the pile-up. The second stage is the fast increase of the bearing area which corresponds to a transition from the pile-up to the bulk material. The final stage is the parabolic increase of the bearing area which results from the change of the intersection area of the quadrangular pyramid shape.

The transition stage is hardly discernible from the material volume ratio curve (Fig. 4.25(b)) and the void volume ratio curve (Fig. 4.25(c)). However, they show a distinctive turning point at the change from truncating the pile-up to the bulk material. Taking the turning point as the level of the original surface plane, some geometric quantities of the indentation (by ignoring the influence of the roughness of indentation topography) are obtained as follows. The pile-up height is 10.1 μm and the indentation depth below the original surface is 49 μm. Thus the pile-up height is about 20% of the indentation depth. The material volume ratio above the original surface plane is 2.63%, while the void volume ratio beneath the original surface plane is 15.75%. Thus the true pile-up volume above the original and the void volume beneath the original surface plane can be derived from them by multiplying the nominal volume which is the total volume (material and void) of the surface topography.[32]

The characterisation procedure introduced here is not only applied to the characterisation of indentation topography; they may be applied to other similar situations where definite micro-topographic quantities are required e.g. lubrication and wear tests.

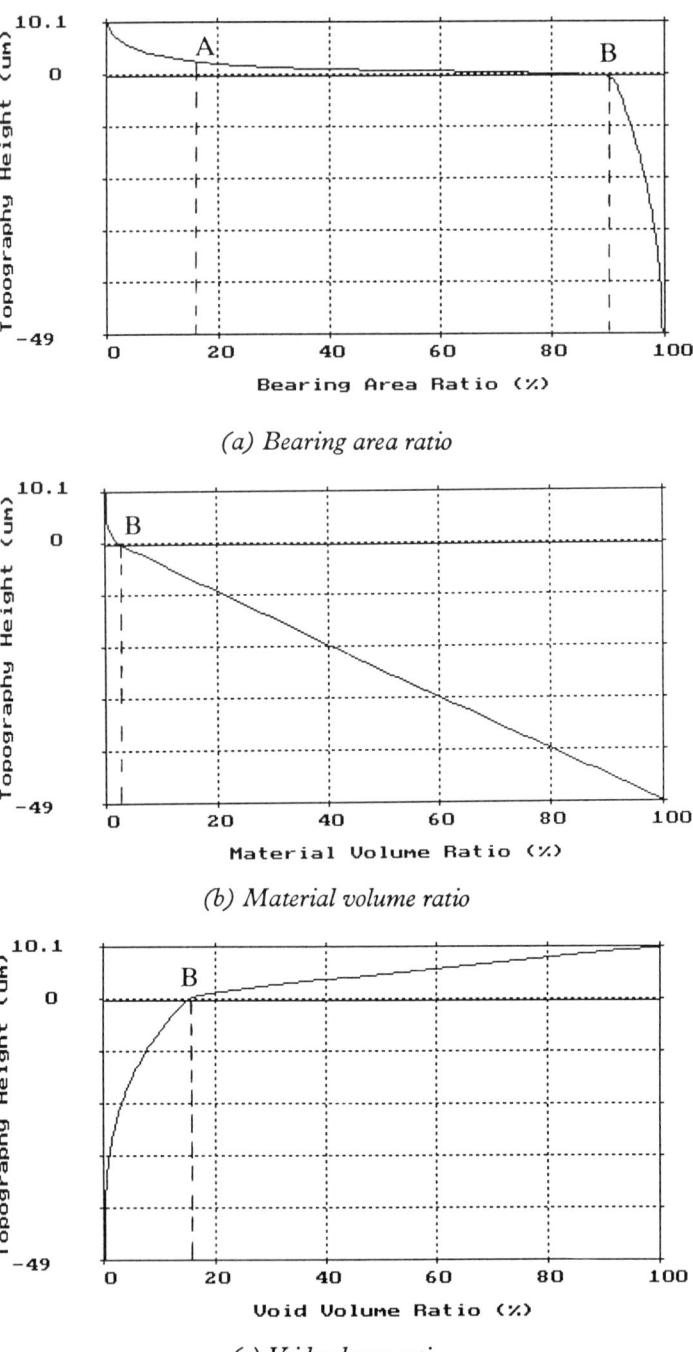

(a) Bearing area ratio

(b) Material volume ratio

(c) Void volume ratio

Fig. 4.25 *Bearing area ratio, material volume ratio and void volume ratio of the indention*

4.9 CONCLUSIONS

In this Part some examples of the applications of 3-D surface metrology have been introduced. Nowadays the practical applications of 3-D surface metrology are being tried and adopted in a variety of areas. It is expected that as surface metrology gains wider exposure in both academic and practical fields more and more applications will be exploited to exhibit its vast potential.

Technical Specification of Some 3D Topography Instruments

1A.1: STYLUS-BASED CONTACTING SYSTEMS

1A.1.1: 3-D Automated Surface Topography Analysis System (ASTA)

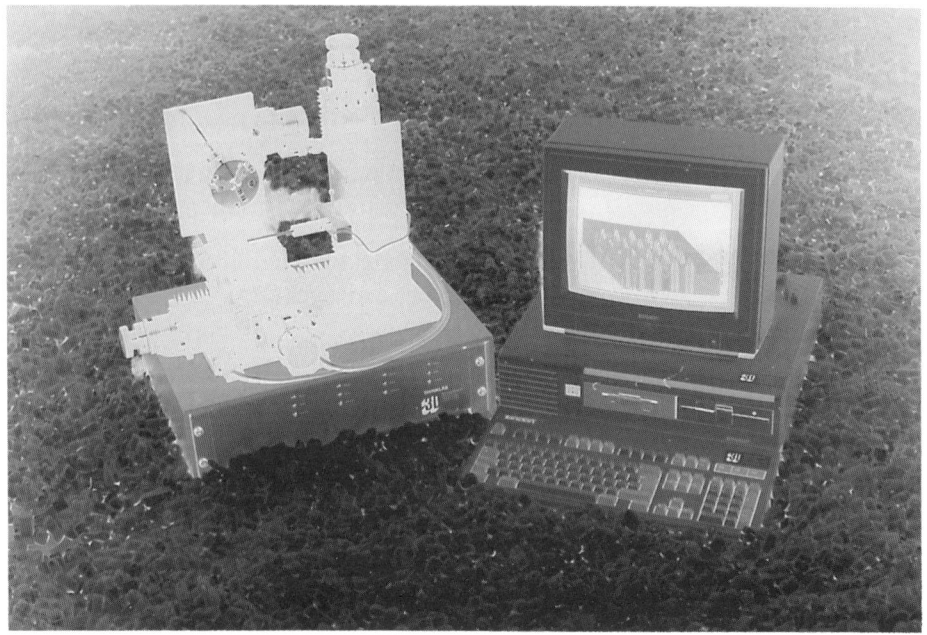

Figure 1A.1: Photograph of the ASTA system (courtesy of 3D Digital Design and Development Ltd, London).

SUPPLIER:	3D Digital Design and Development Ltd.
SCOPE:	Hardware and software.
CONFIGURATION:	3 tables (x, y and z) + rotary stage and stepper motors (see Figure 1A.1).
QUANTISATION:	16 bit.
SAMPLING:	User-determined.
MAX. TRACES:	No limit.
POINTS PER TRACE:	32537 max.
RESOLUTION:	1.25 μm or 0.65 μm. (Linear) and 0.01 degree. (Rotary table).
SOFTWARE:	Interactive and menu-driven. Calibration, Set-up, Analysis, Data presentation, File management, Data import and export.
DISPLAY:	Mainly axonometric also data manipulation.
ANALYSIS:	Wear volume calculations, Roughness (RMS), Histogram of surface distributions.
PRINT OPTION:	To 9/24 pin dot matrix or digital plotter (only axonometric plots or traces).
FILE FORMAT:	Binary, ASCII or Lotus 1-2-3.
ADDRESS:	3D Digital Design and Development Ltd. Interface House Chelmsford Southgate London N14 4JN UK Tel: +44 (81)-886 3668 Fax: +44 (81)-882 4615.

1A.1.2: Perthometer S8P

Figure 1A.2: Photograph of the Perthometer S8P (courtesy of Feinprüf-Perthen GmbH, Göttingen).

SUPPLIER:	Feinprüf Perthen GmbH.
SCOPE:	Hardware and software.
CONFIGURATION:	2 orthogonally-mounted tables (x,z) and gearbox-actuated y-axis (see Figure 1A.2).
QUANTISATION:	14 bits.
SAMPLING:	8000 values per trace (fixed).

TRACES:	From 4 to 129.
ASSESSMENT LENGTHS:	5 Fixed, some selectable. From 0.4 mm to 40 mm.
CUT-OFF:	0.025, 0.08, 0.25, 0.8, 2.5, 8.0 (mm).
Z-RANGE:	Up to 2.5 mm (depending on stylus)
DATA TYPES:	Unfiltered, Roughness, and Waviness.
FILTER TYPE:	Digital, Gaussian (DIN 4777), RC & M Filter.
SOFTWARE:	Interactive and menu driven. Abbott-Firestone curves for traces, Amplitude density curves, 3-D axonometric plots with limited rotation. A vast range of 2-D parameters; no 3-D parameters.
OUTPUT:	Graphics printer, 8 dots per mm. Print-out on light-sensitive paper.
OPTIONS:	Optical probe FOCODYN permits non-contact measurements.
INTERFACE:	2 x RS232 ports for data interchange.
ADDRESS:	Feinprüf Perthen GmbH. P.O. Box 11853 D-37008 Göttingen Germany. Tel: +49 (551) 7073-0 Telex: 96845 Fax: +49 (551) 71021.

1A.1.3: Surfcom 475/575-3D

MANUFACTURER:	Tokyo Seimitsu Co. Ltd.
MAGNIFICATION:	maximum – x10000; minimum – x100 (z-axis). x1, x2, x5, x10, x20,..x100, x200 (horizontal).
TRACE SPEED:	minimum: 0.03 mm/s. maximum: 3 mm/s. return: 3 mm/s.
STRAIGHTNESS X:	$(0.05+1.5L/1000)$ µm.
Y:	$(0.05+3L/1000)$ µm.
RECORDING AREA:	370(X) x 250(Y) mm.
TABLE AREA	80mm x 120mm

I: Surfcom 475–3D

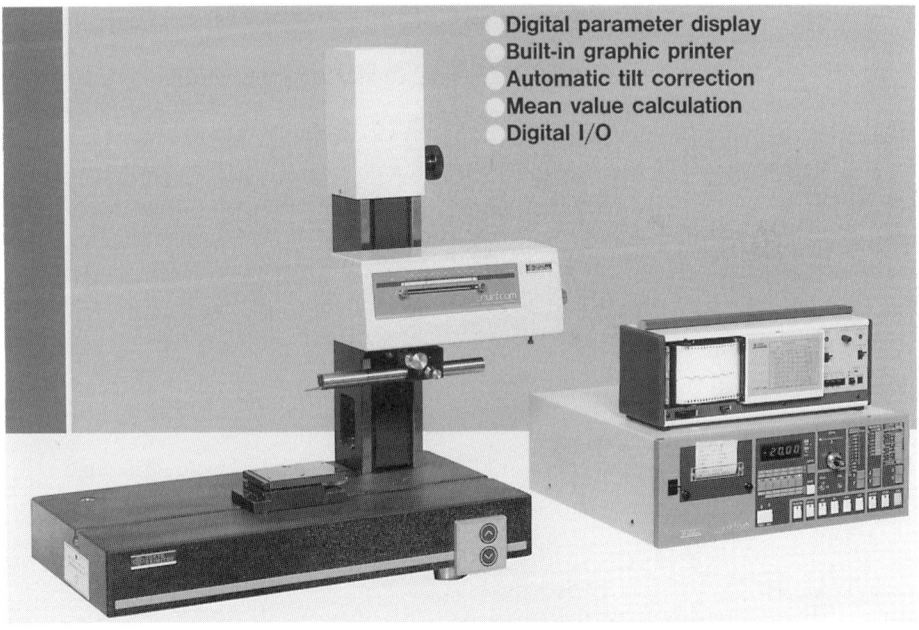

Figure 1A.3: Photograph of the Surfcom 475-3D (courtesy of Advanced Metrology Systems, Leicester).

PARAMETERS: 11 roughness.
                     6 waviness.

DISPLAY: Digital (see Figure 1A.3).

CUT-OFF: 0.25 to 2.5 mm.

FILTER: Analogue (JIS-rated).

LEVELLING: Least squares mean line.

II: Surfcom 575-3D

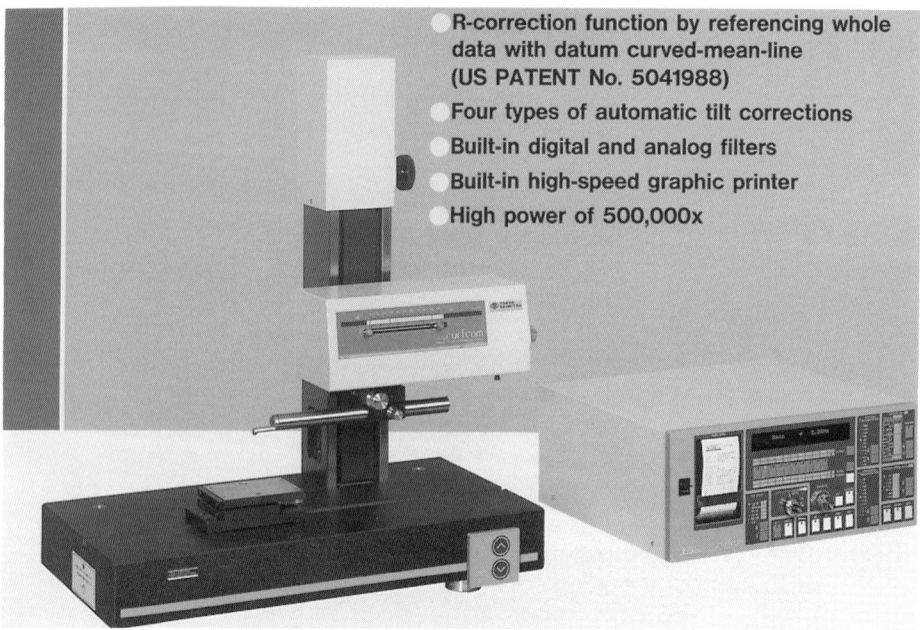

Figure 1A.4: Photograph of the Surfcom 575-3D (courtesy of Advanced Metrology Systems, Leicester).

PARAMETERS: 26 roughness.
                     6 waviness.

DISPLAY: Digital (see Figure 1A.4)

CUT-OFF:	0.025 to 25 mm (Roughness).
FILTER:	Analogue and phase-corrected digital filter.
LEVELLING:	Least squares mean line, curved mean line, beginning half level, end half level both ends level.
PRINTING:	Built-in high-speed graphics printer.
INTERFACE:	Can be connected to external data processing equipment via the RS-232C.
STYLUS FORCE:	4 mN or less.
CONTACT:	Tokyo Seimitsu Co. Ltd. 9–7–1 Shimorenjiaku Mitaka-Shi Tokyo 181 Japan. Tel: 81(Japan)422-48-1019 Fax: 81(Japan)422-49-7315.
AGENT:	Advanced Metrology Systems Ltd. 2 Pomeroy Drive Oadby Industrial Estate Oadby Leicester LE2 5NE UK Tel: (+44) 533 719531 Fax: (+44) 533 720638

1A.1.4. Surfascan 3-D

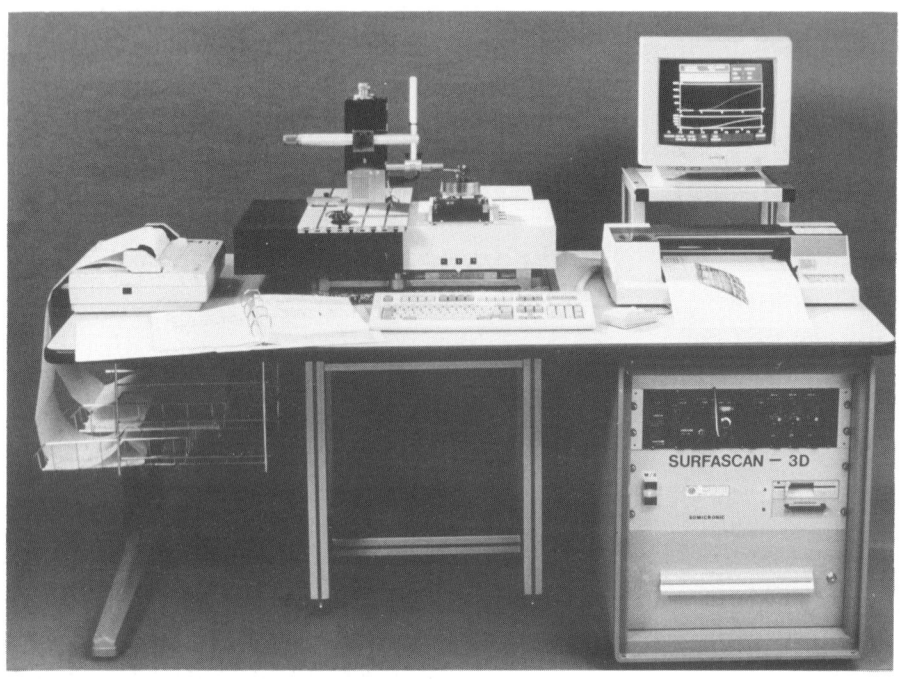

Figure 1A.5: Photograph of the Surfascan-3D (courtesy of Somicronic, St. André de Corcy).

MANUFACTURER: Somicronic.

SCOPE: Hardware and software.
                 ISO, NF, and DIN 2-D parameters.
                 System mounted on anti-vibration stand to
                 isolate measurement unit from external
                 vibrations.

CONFIGURATION: x, y, z translational axes (see Figure A1.5).

QUANTISATION: Carried out by a specific process independent of
                 ADC circuit.

SAMPLING: User-determined.

RESOLUTION: Up to 5 nm vertical.

SOFTWARE: Interactive and menu-driven.
Multiple zooms, ISO, DIN and NF 2-D parameters.
2-D form removal for surface texture analysis.
2-D form characterisation.
Possibility of inversion or rotation, contour plots, 3D bearing curves and 3D filtering.

SET-UP: Measurement of large components (up to 80 kg) possible by changing stylus set-up (see Figure 1A.6). Automatic positioning of probe in the measurement space.

STYLUS: 10 μm radius, 90 degree angle.
2 mN stylus load.
8 different types of interchangeable styli with sizes down to 2.5 μm.

Figure 1A.6: Measurement of a cam shaft using the Surfascan-3D (courtesy of Somicronic, St. André de Corcy).

FILE FORMAT: Binary

RANGE: up to 4 mm (vertical)
                 90 x 90 mm (horizontal)

HARDARE DATUM: Ceramic plane reference surface external to the
                 workpiece and common to both axes.

OPTIONS: Optical probe possible for non-contact
                 measurements.

OUTPUT: Colour graphical display on screen as well as on
                 colour printer (A4). Graphical display on laser
                 printer.

ADDRESS: Somicronic
                 ZI la Vernangère
                 01390 St. André de Corcy
                 France
                 Tel: (+33) 72 26 19 82
                 Fax: (+33) 72 26 19 70

DISTRIBUTOR: UK: Whitestone Business Communications
                 24 Golf Drive
                 Whitestone
                 Nuneaton
                 Warwickshire CV11 6LY
                 UK
                 Tel: +44 (203) 387571
                 Fax: +44 (203) 387571

1A.1.5. Form Talysurf Series

Figure 1A.7: Photograph of the Form Talysurf (courtesy of Rank Taylor Hobson Ltd., Leicester).

SUPPLIER:	Rank Taylor Hobson Ltd.
SCOPE:	Hardware and software for simultaneously measuring waviness, form and roughness.
CONFIGURATION:	y-axis table and gearbox-actuated stylus (x-axis), with interferometrically measured z-component (see Figure 1A.7).
QUANTISATION:	20 bits approx. (interferometric equivalent). 15 bits (inductive).
RANGE:	6 mm (z-axis for interferometric). 1 mm (z-axis for inductive systems).
RANGE/ RESOLUTION:	600 000 (interferometric) 32 768 (inductive)

RESOLUTION:
10 nm (interferometric)
30 nm (inductive with 1 mm range)
0.25μm or 1 μm (horizontal) depending on the traverse length.

TRAVERSE:
120 mm (interferometric)
50–120 mm (others)

SPEED:
10 mm/s (maximum)
1.0 mm/s and 0.5 mm/s measurement speeds.
2.5 mm/s return – stylus on surface.
5.0 mm/s return – stylus off surface.

FORCE:
1 mN nominal (interferometric)
1 mN nominal (inductive)

STRAIGHTNESS:
0.5 μm over 120 mm,
0.1 μm over any 20 mm.
depends on traverse unit.

SOFTWARE:
Interactive, versatile and menu-driven.
Capable of axonometric, contour and intensity plots. Also available are manipulative routines such as zoom, truncation, rotation, inversion and a host of 2-D parameters, as well as trace operations on any traces selected from the surface.

FILE FORMAT:
Binary, with ASCII converter provided.

OPTIONS:
Different types of pick-ups available, e.g. contour pick-up and wide range pick-up with measuring ranges of 20 mm and 28 mm, respectively, and employing 15 bit digitisation.

ADDRESS:
Rank Taylor Hobson Ltd.
P.O. Box 36
New Star Road
Leicester LE4 9JQ
UK
Tel: +44 (533) 763 771
Fax: +44 (533) 741 350

1A.2: OPTICAL AND OTHER NON-CONTACTING SYSTEMS

1A.2.1. Micromap 512 Optical Profiler

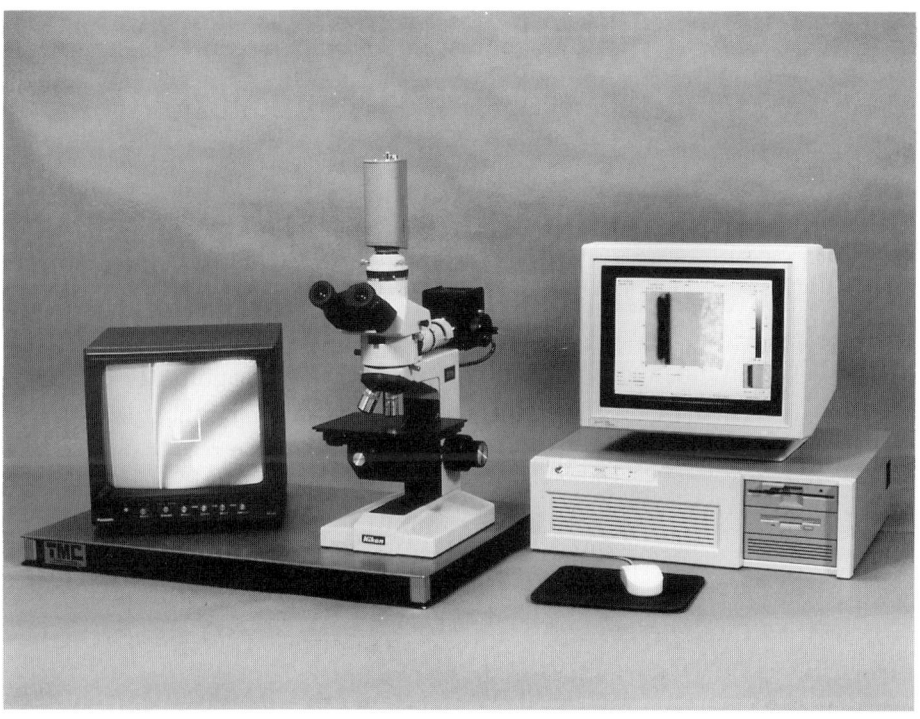

Figure 1A.8: Photograph of the Micromap 512 Optical Profiler (courtesy of Burleigh Instruments (UK) Ltd, Harpenden).

MANUFACTURER:	Micromap Corporation
SCOPE:	Quantitative and visual
PRINCIPLE:	Phase measurement microscopy using interferometry; monochromatic mode for samples with sub-micron roughness; white light mode for roughness with step measurements above 1 mm (picture in Figure 1A.8).
RESOLUTION:	0.2 ångströms (z – at all magnifications) 0.55 µm at x40 magnification

REPEATABILITY:	Less than 1 ångström.
WAVELENGTH:	450 nm to 650 nm.
STAGE TRAVEL:	102 mm(x) by 86 mm (y) with 3° tilt. Optional microscope and stages for samples up to 1 m.
MICROSCOPE:	Compatible with Nikon Optiphot-M stand, CFDI and CFTI objectives.
DATA ARRAY:	600 x 472
SAMPLING INTERVAL:	From 0.34 to 5.4 µm depending on magnification.
SLOPE:	Tolerates a maximum slope of 22° at x40 magnification.
DEPTH OF FOCUS:	Maximum of 50 µm at x2.5 magnification Minimum of 1.1 µm at x40 magnification
MAGNIFICATIONS:	from x2.5 to x40
REFLECTIVITY:	Requires a minimum sample reflectivity of 1%.
SAMPLING AREA:	From 2.7 x 2.5 mm to 170 x 160 µm depending on magnification
CAMERA:	13.5 µm or 7.4 µm pixels, square sampling.
COMPUTER:	AT-compatible 80386 or 80468.
SOFTWARE:	Mouse-driven. Possibility of contour maps, and oblique plots on a 256 colour monitor with 64 grey levels. Also pop-up menus and interactive cursor controls data inspection and analysis.
PARAMETERS:	Roughness statistics such as RMS and peak-to-valley parameters and other values such as R_a, R_{tm}, R_q, R_p, etc. Step and volume calculations.
PRINTOUT:	Includes full colour support for HP Paintjet and video printers.

APPLICATIONS: Applications in measurement of precision
                 surfaces and surface finish.

ADDRESS: Micromap Corporation
                 3131 N Country
                 Club Road, Suite 106
                 Tucson
                 AZ 85716, USA
                 Tel: +1 (602) 881-1911
                 Fax: +1 (602) 881-1913

DISTRIBUTORS: UK: Burleigh Instruments (UK) Ltd
                 Nine Allied Business Centre
                 Coldharbour Lane
                 Harpenden
                 Herts AL5 4UT
                 UK
                 Tel: +44 (582) 766888
                 Fax: +44 (582) 767888

     Germany: Atos GmbH
                 Bergstrasse 104–106
                 D-64319 Pfungstadt
                 Germany
                 Tel: (+49) 6157-3081
                 Fax: (+49) 6157-85990

     France: Eotech SARL
                 17 Rue Gutenburg
                 ZI de la Butte
                 91620 Nozay
                 France
                 Tel: (+33) 1-64 49 71 30
                 Fax: (+33) 1-64 49 32 29

1A.2.2. Wyko Topo-3D

Figure 1A.9: Photograph of the Wyko TOPO-3D (courtesy of Wyko Corporation, Tucson, Arizona).

SUPPLIER:	Wyko Corporation.
SCOPE:	Quantitative and visual (see Figure 1A.9).
PRINCIPLE:	Use of white light in phase measurement optical interferometry.
INTERFEROMETER:	Computer controlled Mirau, Michelson, or Linnik interferometers.
ACTUATION:	Actuation of the stages is by the use of the piezoelectric translator (PZT).
SPEED:	Measurement completed in seconds.
REPEATABILITY:	Measurements repeatable to 0.3 nm RMS.
DETECTOR ARRAY:	256 x 256 photodiode.

DISPLAY
RESOLUTION: 512 x 400.

DIGITISATION: 10 bits.

RANGE: From 0.3 nm to 15 µm (z-axis).

PIXEL SPACING: 40 µm.

SOFTWARE: Fairly comprehensive with full system control, surface calculation, graphic display and statistical analysis. Visual representations include axonometric, intensity plots, histograms, bearing ratio and profiles. Also available are slope and curvature analyses and power spectral analyses.

PARAMETERS: Only a few parameters – RMS, roughness average and peak-to-valley parameters are available.

INTERFACE: TOPO data can be transferred to other computers via the RS-232.

COMPUTER: Uses Hewlett-Packard Series 300 desktop computers. These are based on the MC68030 32-bit microprocessor. The processor operates at a clock speed of 25 MHz and each system has a 10 MB hard disk drive.

APPLICATIONS: Super-polished optics, transparent film surfaces, ball bearings, lens molds, fibre optics, magnetic media and heads and other optical heads.

ADDRESS: Wyko Corporation
                          2650 E. Elvira Road
                          Tucson,
                          AZ 85706, USA
                          Tel: (602) 741-1044
                          Fax: (602) 294-1799

1A.2.3. Wyko RST

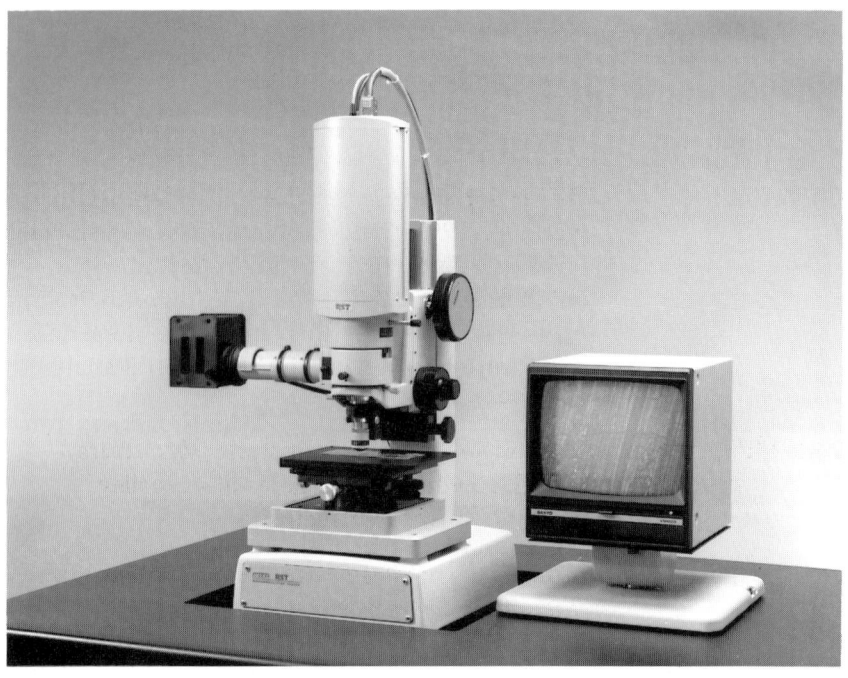

Figure 1A.10: Photograph of the Wyko RST (courtesy of Wyko Corporation, Tucson, Arizona).

SUPPLIER: Wyko Corporation.

SCOPE: Quantitative and visual (see Figure 1A.10).

PRINCIPLE: Uses white light in vertical scanning interference microscopy.

LIGHT SOURCE: Tungsten-halogen lamp.

STAGES: Tip/tilt (6); X/Y translation (50 mm); rotation (90).

SPEED: Measurement completed in seconds.

DETECTOR ARRAY: 248 x 239 pixels.

CAMERA: Solid-state CCD array.

RANGE: from 0.1 nm to 100 μm (z-axis); lateral range
                    depends on magnification.

RESOLUTION: 3 nm vertical. Lateral resolution depends on the
                    objective used; 7.7 μm for x2.5 and 0.48 μm for
                    x40.

SOFTWARE: Includes surface calculation, graphic display and
                    statistical analysis. Visual representations include
                    axonometric, contour plots, histograms, bearing
                    ratio and profiles. Also a range of analysis
                    enhancement features including inversion,
                    rotation, subsection selection, and expansion.
                    Calculated parameters can be stored in a
                    database.

PARAMETERS: A range of common statistical parameters –
                    RMS, roughness average, peak-to-valley
                    parameters etc. are available.

WORK-TOP: Optional workstation isolation table optimises
                    system setup and use.

PC REQUIREMENTS: It uses a 486 PC-compatible computer system
                    which makes it easy to interface to local area
                    networks (LANs).

COMPONENTS: Includes microscope, video display, an
                    interferometric microscope objective, a 486
                    computer with keyboard and VGA monitor,
                    calibration standard and software.

APPLICATIONS: Paper, rolled steel, sheet metal, laser-engraved
                    rollers, plastics, painted surfaces, diamond films,
                    wafers and ceramics.

ADDRESS: Wyko Corporation
                    2650 E. Elvira Road
                    Tucson,
                    AZ 85706, USA
                    Tel: (602) 741-1297
                    Fax: (602) 294-1799

1A.2.4. UBM Optical Surface Measurement System

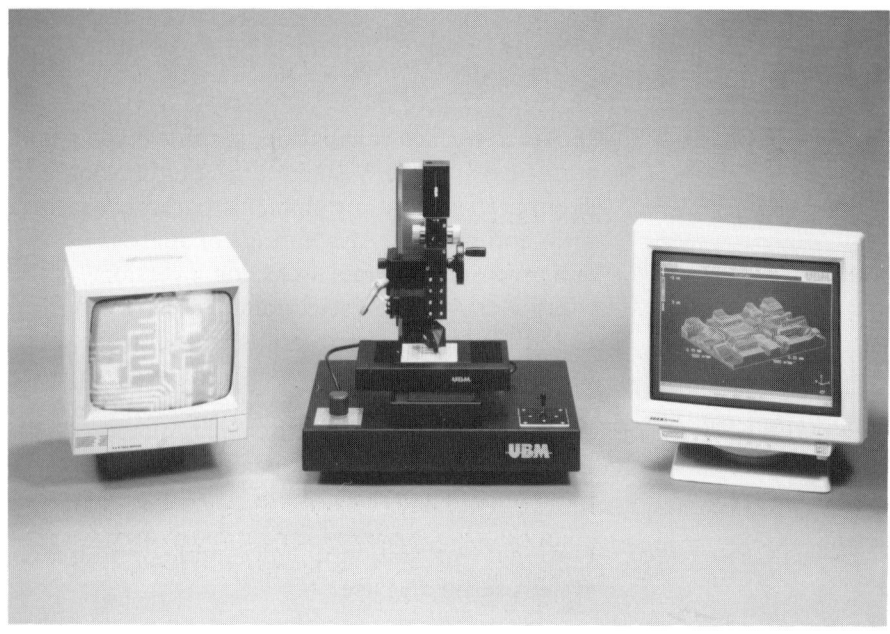

Figure 1A.11: Photograph of the UBM Optical Surface Measurement System (courtesy of UBM Messtechnik GmbH, Ettlingen).

SUPPLIER: UBM Messtechnik GmbH.

SCOPE: Quantitative and visual (see Figure 1A.11).

PRINCIPLE: Laser dynamic focusing principle.

SENSORS: A wide variety of sensors are available ranging from the Laserfocus through the Superfocus to the Nanofocus which is ideal for very high frequency vibration amplitude measurement. Also available is a rotationally symmetric triangulation sensor for a greater range at somewhat lower resolution.

ACTUATION: Actuation of the stages is by the use of stepping
                        or DC motors with either mechanical or air
                        bearings depending on the demands of the
                        application.

SPEED: Measurement completed in minutes.
                        Maximum translational velocity of 10 mm/s.

DISPLAY: VGA resolution with 255 colours.

DIGITISATION: 14 bits.

RANGE: Depends on the sensor used. Varies from 1 μm
                        (Nanofocus) to 20 mm for the triangulation
                        sensor. The dynamic focusing sensors have two
                        ranges – 100 μm and 1 mm. Horizontal range is
                        50–300 mm depending on translational stage
                        option.

RESOLUTION: 1 μm (triangulation) and 0.01 μm or 0.1 μm
                        (depending on selected range) for dynamic
                        focussing sensors. The Nanofocus has a
                        resolution of 1 nm.

REPEATABILITY: Positional repeatability is typically 1 μm for
                        mechanical bearing systems and 0.5 μm for air
                        bearing systems.

SPOT DIAMETER: For dynamic focusing sensors 1 μm. For
                        triangulation sensors 50 μm.

SURFACE SLOPE: ±20° for most systems on perfectly mirrored
                        surfaces; higher on rough surfaces.

SOFTWARE: Comprehensive with full system control, surface
                        calculation, graphic display and statistical
                        analysis. Up to 15 graphic images may be
                        presented per page or a batch sequence printed
                        overnight. Included is the possibility of creating
                        high-quality photographs and slides for
                        documentation or other uses. Software is used to
                        correct drift which occurs when measurements
                        are taken over a long period of time. Also
                        available are step height measurement, zoom,
                        rotation, levelling, angle measurement, filtering,

averaging, fractal dimension, 2D and 3D roughness parameters, polynomial fitting, 2D and 3D FFT, power spectrum, auto and cross-correlation, contour and bearing plots. The measurement and analysis can be automated using Macro sequences. The customer logo can be included in screen displays and other statistical software packages can be easily integrated.

PARAMETERS: Many 3D parameters including – RMS, roughness average and peak-to-valley parameters are available. Also fractal dimension and slope distribution.

PC REQUIREMENTS: IBM compatible AT 386, 40 MHz, VGA display, MS DOS, 85 MB hard disk drive, 3.5-inch 1.44 MB floppy disk drive, 2 MB RAM.

APPLICATIONS: Wet paste, IC lead planarity measurement, micro-mechanical elements, flatness measurement of relay contacts, silicon membrane thickness measurement, profile and roughness measurement of paper and plastics, profilometry of skin, glass and contact lenses etc.

ADDRESS: UBM Messtechnik GmbH
Ottostrasse 2
D-76275 Ettlingen
Germany
Tel: +49 7243 16025
Fax: +49 7243 17583

1A.2.5. Feinprüf Perthen RM-600 3-D

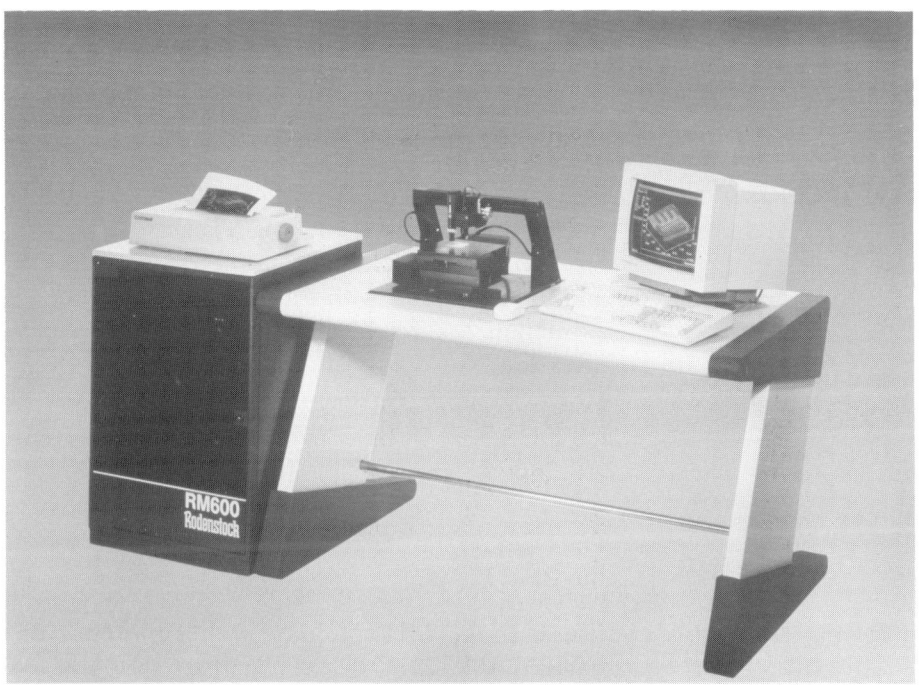

Figure 1A.12: Photograph of the Feinprüf Perthen RM-600 (courtesy of Feinprüf-Perthen GmbH, Göttingen).

SUPPLIER:	Feinprüf Perthen GmbH
SCOPE:	Quantitative and visual (see Figure 1A.12 or Figure 4.9).
PRINCIPLE:	Use of laser in focus-detect non-contact measurement.
ACTUATION:	Low vibration precision tables (2 μm resolution) with mechanical stops for relative positioning.
SPEED:	Measurement completed in minutes.
SPOT SIZE:	1 or 2 μm in diameter (spatial resolution, 2 μm).

RESOLUTION: ±0.2 μm (vertical) or 0.1% independent of range.

MIN.   2 per cent reflectivity of the surface is adequate
REFLECTIVITY: for valid measurements.

DIGITISATION: 10 bits.

RANGE: From 0.02 μm to 600 μm (z-axis).

SOFTWARE: Full system control, surface calculation, graphic
                     display and statistical analysis. Visual
                     representations include axonometric, contour
                     and intensity plots with zoom, levelling, profile
                     extraction, scaling, filtering, rotation and
                     inversion.

PARAMETERS: Measurement of distance, angle, circle fits (for
                     radius calculation), height, area plus a range of
                     2-D roughness values.

INTERFACE: Data can be transferred to other computers via
                     the RS-232.

PC REQUIREMENTS: It uses an AT/40H computer with a minimum of
                     4 MB RAM.

APPLICATIONS: Super-polished optics, transparent film surfaces,
                     ball bearings, lens moulds, fibre optics, magnetic
                     media and heads and other optical heads.

OPTIONAL: Air bearing translational (X,Y) table available.
                     Positioning microscope also available. Also video
                     camera with monitor.

ADDRESS: Feinprüf Perthen GmbH
                     P.O. Box 11853
                     D-37008 Göttingen
                     Germany
                     Tel: +49 (551) 7073-0
                     Telex: 96845
                     Fax: +49 (551) 71021

1A.2.6. Scanning Force Microscope (SFM)

SUPPLIER: Centre Suisse d'Électronique et de
                     Microtechnique S.A.

SCOPE: Visual only – no statistical parameters.

PRINCIPLE: Use of scanning a sharp cantilever stylus over a
                     sample and measuring deflection.

MODES: Two modes – contact and non-contact.

SPEED: Measurement completed in seconds.

ACTUATION: Three axis PZT micro slide.

RESOLUTION: Atomic resolution (2 ångström) lateral.
                     0.1 Å vertical.

RANGE: 1 µm and 3 µm (z-axis of scanner A and B).
                     1 µm and 15 µm lateral – scanner A and B.

SOFTWARE: Mainly 3-D axonometric display.

PARAMETERS: No statistical parameters.

FLEXIBILITY: Retrofitable to existing STM system.
                     Operational in vacuum, air and liquid.

APPLICATIONS: Conductive and non conductive samples,
                     semiconductors, optical fibres), biological and
                     other very sensitive samples, *in situ* analysis of
                     samples immersed in liquids.

ADDRESS: CSEM
                     Rue Breguet 2
                     CH-2007 Neuchâtel
                     Switzerland
                     Tel: (+41) 38 205 111
                     Fax: (+41) 38 205 640

1A.2.7. Maxim 3-D Model 5700

Figure 1A.13: Photograph of the Maxim 3-D Model 5700 (courtesy of Zygo Corporation, Middlefield, CT).

SUPPLIER:	Zygo Corporation and Lambda Photometrics Ltd.
SCOPE:	Quantitative and visual (Figure 1A.13).
PRINCIPLE:	Use of low power laser in interferometry together with optical microscopy.
LIGHT:	0.6328 μm He-Ne laser.

INTERFEROMETER: MicroFizeau and Mirau interferometer objectives.

TRAVEL: Motorised or manual stage with a travel of
                  152 mm in all 3 axes.

SPEED: Measurement completed in less than 5 seconds.

REPEATABILITY: Measurements repeatable to 0.5 nm RMS.

DETECTOR ARRAY: 256 x 256 photodiode.

DIGITISATION: 10 bits.

RANGE: 40 µm (z-axis).

AREA: From 70 µm to 7 mm square.

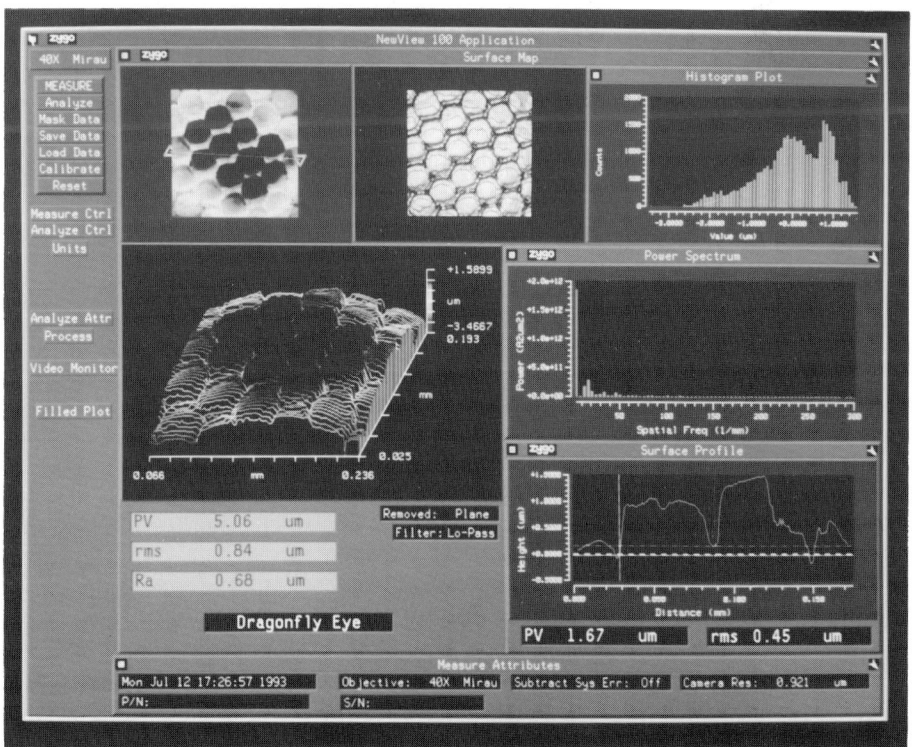

Figure 1A.14: Screen display of the Maxim 3-D Model 5700 topography instrument (courtesy of Zygo Corporation, Middlefield, CT).

RESOLUTION: 0.5 ångström (vertical).
                     0.36 to 26.4 µm (lateral).

PIXEL SPACING: 28 µm.

REFLECTIVITY: 4 per cent (less than 4% available).

SOFTWARE: Fairly comprehensive with full system control,
                     surface calculation, graphic display and statistical
                     analysis. Visual representations include
                     axonometric, intensity plots, histograms, bearing
                     ratio and profiles (see Figure 1A.14). It uses
                     windows and icons and is relatively easy to use.
                     Customised versions are also available.

PARAMETERS: A wide variety of surface characterisation
                     parameters – RMS, roughness average and
                     peak-to-valley parameters are available. Also
                     included are angle, twist tilt measures as well as
                     spectral analyses.

APPLICATIONS: Coated and uncoated metals, glass, ceramics,
                     plastics and other materials with specular
                     surfaces.

ADDRESS: Zygo Corporation
                     Laurel Brook Road
                     P.O. Box 448
                     Middlefield
                     CT 06455-0448, USA
                     Tel: (203) 347-8506
                     Fax: (203) 347-8372

1A.2.8. MP2000

SUPPLIER:	Chapman Instruments.
SCOPE:	Quantitative and visual (see Figure 1A.15).
PRINCIPLE:	Use of laser in slope detection.
INTERFEROMETER:	Nomarsky interferometers.
SPEED:	Measurement completed in seconds (2 mm/s).
REPEATABILITY:	Measurements repeatable to nm RMS.

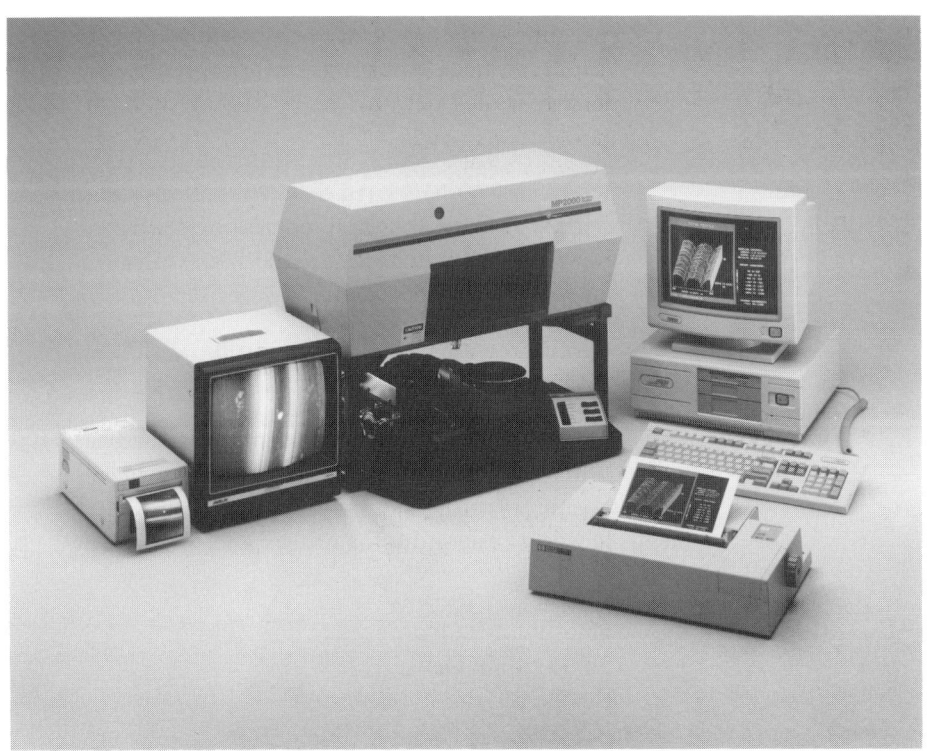

Figure 1A.15: Photograph of the MP2000 (courtesy of Chapman Instruments, Rochester, NY).

RESOLUTION: Sub-ångström (vertical height resolution) high resolution encoder permits better spatial resolution of 0.2 μm.

RANGE: tens of microns (z-axis).

SCAN LENGTH: up to 100 mm in 2D, 100mm x 100 mm in 3D.

SOFTWARE: Allows for the automatic display of key surface waveforms such as waviness, roughness, total profile or slope as well as associated surface parameters. Displays include axonometric and contour together with zooms, and overlays.

PARAMETERS: Over 30 parameters available including – RMS, roughness average and peak-to-valley parameters are available; also histograms, power spectra and more.

PC REQUIREMENTS: Features 486 processor running at 33 MHz as well as a 240 MB hard disk drive. It also comes with a 15-inch 16-bit video graphics adaptor board to provide high speed, high resolution colour enhanced graphics.

OPTIONS: Circumferential (orbital) scanning, disk and wafer handling, Nomarsky viewing system.

APPLICATIONS: Optics, hard disks, magnetic materials, paper, films and coated materials, ceramic surfaces, aspheres, mirrors, intraoccular and contact lenses, semiconductor wafers and flat panel displays.

ADDRESS: Chapman Instruments
50 Saginaw Drive
Rochester
NY 14623, USA
Tel: (716) 461-1950
Fax: (716) 461-0517

1A.2.9. Proscan 1000

SUPPLIER: Scantron Ltd.

SCOPE: Quantitative and visual (see Figure 1A.16).

PRINCIPLE: Use of laser-based displacement probes for
                       non-contact measurement.

SPEED: Measurement completed in minutes.

REPEATABILITY: From ±0.01 µm to ±6.0 µm depending on spot
                       size and measuring range.

SPOT SIZE: From 10 µm to 300 µm depending on the probe
                       selected (six different probes available).

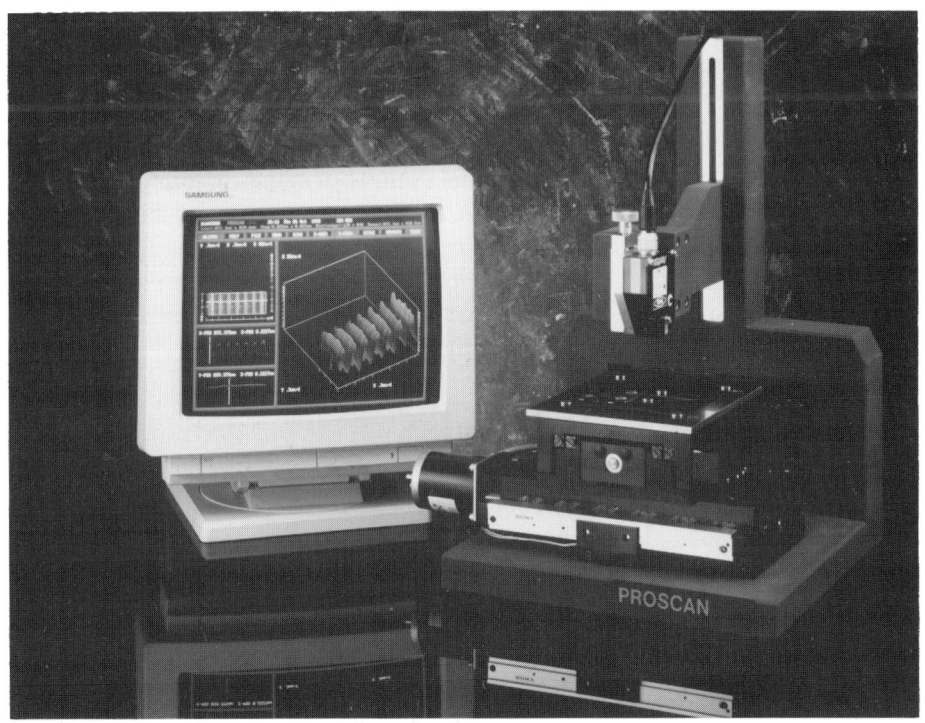

Figure 1A.16: Photograph of the Proscan 1000 (courtesy of Scantron Ltd, Somerset)

RANGE:	From hundreds of microns to hundreds of millimeters depending on the probe selected.
MIN. STEP:	1 micron (X, Y); optional 0.1 micron step available.
MAX. STEP:	9.999 mm.
MAX. NO. DATA POINTS:	160,000/part. Optional – 1,000,000.
MAX. DATA RATE:	2000 points per second.
SOFTWARE:	Easy to use dedicated software with pull down menus. Uses super-high resolution VGA screen, 120 colour shades with four display windows. Has the following display types – (1) 3-D isometric view of scanned component with relative height shown in colour bands, (2) intensity plots, (3) cross-section profile of x-axis, and (4) cross-section profile of y-axis. Also available are selectable scaling of z-measurement, zoom, manual control of stages for individual z-axis measurements, nomination of datum position for relative measurements, real-time display of coordinates, facility of comparison of measurement with master sample, search facility for identifying surface irregularities, measurement of length, angle, range, and FFT. Optional CCD camera available for visual alignment of parts.
PARAMETERS:	Only a few parameters – roughness average, length, angle. Volume, cross-section, surface area calculation.
OPTIONS:	CCD camera, high resolution stages, laser probes (six available).
APPLICATIONS:	Solder paste profile, flatness of mechanical seals, digitising of moulds and models, flatness of optical discs, surface topography of delicate materials.

ADDRESS: Scantron LTD
Bindon Industrial Estate
Frobisher Way
Taunton
Somerset TA2 6BB
UK
Tel: (+44) 823 333343
Fax: (+44) 824 333684

STATISTICAL ANALYSIS OF SURFACE TOPOGRAPHY USERS' QUESTIONNAIRE

I. Questionnaires sent out.

Total:		80
Establishment:	Industries	62.5%
	Universities	25%
	Institutes	12.5%
Total		**100%**
Countries:	Germany	30%
	UK	21%
	France	12.5%
	Sweden	10%
	Netherlands	9%
	Italy	5%
	Ireland	4%
	Belgium	4%
	Denmark	2.5%
	Austria	1%
	Switzerland	1%
Total		**100%**

II. Completed questionnaires returned.

Total:		36
Establishment:	Industries	55.5%
	Universities	28%
	Institutes	16.5%
Total		**100%**
Countries:	UK	28%
	France	25%
	Germany	22%
	Sweden	8%
	Belgium	8%
	Netherlands	6%
	Denmark	3%
Total		**100%**

III. Reply analysis

General

1. Please state the environment in which your 2-D or 3-D (Please indicate) systems are used.

    2-D   a) Research 46.5%   c) Industry 42%   c) Other 11.5%

    3-D   a) Research 54%   c) Industry 38%   c) Other 8%

2. Do you feel 3-D surface analysis is important for your work ?

    Yes 89%,   No 3%,   Maybe 8%,   No idea ___,   Other ___

3. What are your main interests in 3-D surface topography ?

    Characterisation 32%, Measurement 32%, Functional properties (Solving Engineering problems) 32%, Others ___

4*. Which 2-D parameters have you got on your system ?
    All parameters from National and International Standards 66.5%, Waviness parameters 11%, Parameters provided by some instruments 22.5%

(*If you do not have the use of a 3-D system please move to question 27.)

Instruments and Measurement

5. Which types of 3-D surface measurement systems do you use ?
    Stylus 45%, Optical 50%, Others (STM,MFM,AFM) 5%

6. If optical, what principles are involved ?
    Interferometric 26%, Scatter 15%, Focus detect 30%, SEM, 11%, Others (Laser stylus, Confocal laser profilometer, Reflection of light, Slope integration, Optical level) 18%

7. How is the measurement motion effected?
    Translational stages 23.5%, Stage & gearbox 63.5%, Other (No motion, Piezo-controlled microscope objective scanner) 13%

8. Please state manufacturer's name of the instrument.
    Rank Taylor Hobson 24%, Perthen, 12% Rodenstock, 8% Hommelwerke 5%, Somicronic 16%, Wyko 5%, Federal 5%, Institute of Optical Research 3%, Heidelberg Instrument 3%, Kontron IBAS 3%, ANASPEL 3%, Carl Zeiss 3%, Flipsesfraat, 3% In-house system 7%

9. Do you use a levelling table ?
    Yes 57%, No 43%

10. Please state the range and resolution in vertical (z-axis) and horizontal (x-, y- axes): Depends on instrument.

    Vertical range: (max. 2mm for stylus, max. 0.5mm for optics)

                        Resolution: (min. 0.005um for stylus, min. 0.2nm for optics)

    Horizontal range: (max. 150mmx150mm)

                        Resolution: (min. 0.1 μm for stylus, 10 nm for optics)

11. How is specimen relocation effected?
Physical fixture 53.5%, Optical 28.5%, Camera 11%, Sensor 7%, Others __

Data Sampling

12. How many bits has your ADC ?
14 17%, 13 9%, 12 48%, 10 17%, 8 9%

13. Please state data sampling method (On-the-fly if data is sampled while there is relative motion between the specimen and the stylus, and static if data is sampled point by point while the stylus and specimen are not in relative motion).
On-the-fly 56%, Static 44%

14. Is the analogue signal filtered prior to analysis?
Yes 35%, No 65%

15. Do you use digital filtering techniques?
Yes 79%, No 21%

16. What are the sampling intervals you adopted most frequently?
Depends on specimen and instrument – from 1 μm for stylus systems and 1 nm for AFMs.

17. What are the sampling areas or sampling points you adopted most frequently?
Depends on specimen and instrument – from 0.5mmx0.5mm.

Characterisation and Parameters

18. What characterisation approach(es) do you use?
Statistical 35%, Spectral 15%, Time Series 8%, Functional 27%, Fractal 8%, Others 7% Motif.

19. Which of the above approaches do you feel industry most needs ?
Statistical 29%, Spectral 9%, Time Series 3%, Functional 41%, Fractal 6% Motif 6%

20. What type of parameters have you used in your application ?
    Statistical <u>29%</u>, Visualisation <u>35%</u>, Functional Others (Spectral) <u>5%</u>

21. What kinds of functional characterisation methods do you use ?
    Confidential <u>8%</u>,   Area bearing ratio <u>38%</u>,   Motif <u>19%</u>,
    Comparison with results of functional tests <u>4%</u>,   Truncation <u>4%</u>,
    Problem dependent <u>4%</u>,   Morphological <u>11%</u>,   Spectral <u>4%</u>,
    Visualisation <u>8%</u>

22. What kind of functional information would you like to get from
    surfaces?
    (a) Surface function, contact state of surface, bearing area
    (b) Oil volume, lubrication property of surface
    (c) Wear, friction, tightness
    (d) Isotropic parameters, lay direction parameters
    (e) Plastic deformation, information for FE (Finite Element) modelling
    (f) Influence on optical usage
    (g) Height and portion of peaks and valley
    (h) The role of roughness and waviness
    (i) Ranges of parameters by which relative motion parts contacted are
        best matched and have the longest life
    (j) Morphological information
    (k) Microgeometrical investigations of tribological surfaces

23. Are your present parameters able to give you this information?
    Yes <u>25%</u>, No <u>62.5%</u>, Partly Yes <u>12.5%</u>

24. What 3-D parameters do your system generate ?
    (a) Parameters correspond to 2-D
    (b) Anisotropy index, anisotropy direction
    (c) Distribution and quantification of motifs
    (d) Bearing area ratio, contact area
    (e) Oil volume, volume of pile up and indent
    (f) Geometric parameters, patterning parameters
    (g) Fractal dimension
    (h) Averaged 2-D parameters on a surface
    (i) Slope and curvature distribution

25. What specific functional information do you believe is obtained from
    them?
    (a) Wear, oil consumption
    (b) Contact information
    (c) Deformation information
    (d) Optical behaviour of surfaces

(e) Morphological

(f) Anisotropy information

(g) Image information

26. How is your datum defined?

Arithmetic mean plane 12%, Least square mean plane 73%, Others (Minimum zone plane, Cylindrical or Curvature regression, Digital filtering) 15%

(to be completed by those without current use of 3-D systems)

27. Do you consider 2-D analysis to be adequate for your application ?

Yes 57% (For surface laser cladding and alloying, for some specific usage and for use in the workshop. Sufficient for mass-production process).

No 43%

28. Do you feel that 3-D analysis has a role to play in the present or future performance of your company/institution?

Yes 85%

No 15% (Because long measurement time is needed)

29. What functional information would you most like to get from a 3-D system?

(a) Ability to predict contact pressure in load

(b) Detecting of peaks, valleys, passes, grooves

(c) Harmonic content of the 1st and 2nd frequency

(d) Imaging quality of optical surface

(e) Surface deformation after laser welding and cladding

(f) Characterisation of deterministic texture

(g) Crater and rim parameters as formed by laser or electronic beam texturing

(h) 2-D power density spectrum and 2-D autocorrelation function

(i) Separation of 3-D roughness from 3-D waviness and waviness from form deviation

(j) Volume extraction

(k) Roughness parameters in all directions

BIBLIOGRAPHY

PART I
INSTRUMENTS AND MEASUREMENT
TECHNIQUES OF 3-DIMENSIONAL
SURFACE TOPOGRAPHY

1. T.R. Thomas, *Rough Surfaces*. Longman Press, London (1982).

2. H. Dagnall, *Exploring Surface Texture* (2nd edition). Rank Taylor Hobson Limited (1986).

3. T.V. Vorburger and E.C. Teague, Optical techniques for on-line measurement of surface topography, *Precision Engineering, Vol. 3*, No. 1, pp. 61–83 (1981).

4. I. Sherrington and E.H. Smith, Modern measurement techniques in surface metrology. Part I: Stylus instruments, electron microscopy and non-optical comparators, *Wear, Vol. 125*, pp. 271–288 (1988).

5. I. Sherrington and E.H. Smith, Modern measurement techniques in surface metrology. Part II: Optical instruments, *Wear, Vol. 125*, pp. 289–308 (1988).

6. M. Francon, *Progress in Microscopy*. Pergamon, Oxford (1961).

7. B.L. Clark, *The Encyclopedia of Microscopy*. Reinhold, New York (1961).

8. J.A. Bennett, The social history of the microscope, *Journal of Microscopy, Vol. 155*, Part 3, pp. 267–280 (1989).

9. T.K. Chinmayanandam, On the specular reflection from rough surfaces, *Physics Review, Vol. 13*, p. 96 (1919).

10. V.E. Cosslett, *Practical Electron Microscopy*. Butterworth, London (1951).

11. M.E. Haine and V.E. Cosslett, *The Electron Microscope*. Spon, London (1961).

12. R.D. Heidenreich, *Fundamentals of Transmission Electron Microscopy*. Wiley, New York (1964).

13. T. Mulvey, The electron microscope: the British contribution, *Journal of Microscopy, Vol. 155*, Part 3, pp. 327–338 (1989).

14. P.B. Hirsch, TEM in materials science – past, present and future, *Journal of Microscopy, Vol. 155*, Part 3, pp. 361–371 (1989).

15. M. Knoll, Charging potential and secondary emission of bodies under electron irradiation, *Z. tech. Phys. Vol. 16*, pp. 467–475 (1935).

16. A.D.G. Stewart and M.A. Snelling, A new scanning electron microscope, *Proceedings of 3rd European Conference on Electron Microscope*, pp. 55–56 (1965).

17. P.R. Thornton, *Scanning Electron Microscopy*. Chapman and Hall, London (1968).

18. D. McMullan, SEM – past, present and future, *Journal of Microscopy, Vol. 155*, Part 3, 373–392 (1989).

19. G. Nomarski, Microinterferometer differentiel à ondes polarisées, *Journal of Physics – Radium., Vol. 16*, pp. 9S–13S (1955).

20. A.J. Hale, *The Interference Microscope*. E. & S. Livingstone, London (1958).

21. S. Tolansky, *Introduction to Interferometers*. Wiley, New York (1973).

22. H. Trumpold, Limits of application of the interference methods for surface measurements, *Proceedings of the Institution of Mechanical Engineers, Vol. 182*, Part 3K, pp. 241–254 (1967–8).

23. J.B.P. Williamson, Microtopography of surfaces, *Proceedings of the Institution of Mechanical Engineers, Vol. 182*, Part 3K, pp. 21–30 (1967–8).

24. J. Peklenik and M. Kubo, A basic study of a three-dimensional assessment of the surface generated in a manufacturing process, *Annals of the CIRP, Vol. 16*, pp. 257–265 (1967–8).

25. M. Kubo and Peklenik, An analysis of micro-geometrical isotropy for random surface structures, *Annals of the CIRP, Vol. 16*, pp. 235–242 (1968).

26. J. Wallach, Surface topography description and measurement, *Proceedings of ASME Winter Annual Meeting*, pp. 1–22, Nov. 16–21 (1969).

27. R.S. Sayles and T.R. Thomas, Mapping a small area of a surface, *Journal of Physics E: Scientific Instruments, Vol. 9*, pp. 855–861 (1976).

28. R.D. Young, Field emission ultramicometer, *Review of Scientific Instruments, Vol. 37*, No. 3, pp. 275–278 (1966).

29. R.D. Young, J. Ward and F. Scire, The topografiner: An instrument for measuring surface microtopography, *Review of Scientific Instruments, Vol. 43*, No. 7, pp. 999–1011 (1972).

30. A. Boyde, A single-stage carbon replica method and some related techniques for the analysis of the electron microscopic image, *Journal of the Royal Microscopic Society, Vol. 86*, pp. 359–370 (1967).

31. A. Boyde, Practical problems and methods in the three-dimensional analysis scanning electron microscope images, In O. Johari and I. Corvin (eds) *Scanning Electron Microscopy*, IIT Research Institute, Chicago, pp. 105–112 (1970).

32. A.R. Dinnis, After-lens deflection and its uses, In O. Johari and I. Corvin (eds) *Scanning Electron Microscopy*, IIT Research Institute, Chicago, pp. 41–48 (1971).

33. P.G.T. Howell and A. Boyde, Comparison of various methods for reducing measurements from stereo-pair scanning electron micrographs to real 3-D data, In O. Johari and I. Corvin (eds) *Scanning Electron Microscopy*, IIT Research Institute, Chicago, pp. 233–240 (1972).

34. O. Johari and A.V. Samudra, Scanning electron microscopy, In R.F. Kane and G.B. Larrabee (eds) *Characterisation of Solid Surfaces*, Plenum Press, New York, pp. 107–131 (1974).

35. RM600 – A new approach to surface measurement (Product information), Optische Werke G. Rodenstock (1991).

36. Focodyn – Optical probe for perthometer non-contact profile acquisition (Product information), Feinprüf Perthen GmbH (1991).

37. UB16 – Precision optical length measurement system (Product information), Ulrich Breitmeier Messtechnik GmbH (1991).

38. LSM320 – The LSM confocal laser scan microscope with CPU 80486 and Windows (Product information), Carl Zeiss (1992).

39. M. Minsky, Microscopy apparatus, United States Patent Office. *Filed* November 7 1957, granted December 19 1961 Patent No. 3013467.

40. M.O. Dupuy, High-precision optical profilometer for the study of micro-geometrical surface defects, *Proceedings of the Institution of Mechanical Engineers, Vol. 182*, Part 3K, pp. 255–259 (1967–8).

41. J. Simon, New noncontact devices for measuring small microdisplacements, *Applied Optics, Vol. 9*, No. 10, pp. 2337–2340 (1970).

42. J.A. Simpson, Use of a microscope as a noncontacting microdisplacement measurement device, *Review of Scientific Instruments, Vol. 42*, pp. 1378–1380 (1971).

43. G. Bouwhuis and J.J.M. Braat, Video disk player optics, *Applied Optics, Vol. 17* (1978).

44. T. Wilson, Imaging properties and applications of scanning optical microscopes, *Applied Physics, Vol. 22*, pp. 119–128 (1980).

45. Y. Fainman, E. Lenz and J. Shamir, Optical profilometer: A new method for high sensitivity and wide dynamic range, *Applied Optics, Vol. 21*, pp. 3200–3208 (1982).

46. M. Dobosz, Optical profilometer: A practical approximate method of analysis, *Applied Optics, Vol. 22*, No. 24, pp. 3983–3987 (1983).

47. M. Dobosz, Accuracy of profile measurements by means of a focused laser beam, *Wear, Vol. 98*, pp. 117–126 (1984).

48. D.Y. Lou, A. Martinez and D. Stanton, Surface profile measurement with a dual beam optical system, *Applied Optics, Vol. 23*, p. 746 (1984).

49. T. Kohno, N. Ozawat, K. Miyamoto and T. Musha, Practical non-contact surface measuring instrument with one nanometre resolution, *Precision Engineering, Vol. 7*, No. 4, pp. 231–232 (1985).

50. D.K. Hamilton and T. Wilson, Three-dimensional surface measurement using the confocal scanning microscope, *Applied Physics B, Vol. 27*, pp. 211–213 (1982).

51. J. Mignot and C. Gorecki, Measurement of surface roughness: comparison between a defect-of-focus optical technique and the classical stylus technique, *Wear, Vol. 87*, pp. 39–49 (1983).

52. T. Kohno, N. Ozawa, K. Miyamoto and T. Musha, High precision optical surface sensor, *Applied optics, Vol. 27*, No. 1, pp. 103–108 (1988).

53. S. Tolansky, Multiple-Beam Interference Microscopy of metals, *Academic* (1970).

54. J.M. Bennett, Measurement of the rms roughness, autocovariance function, and other statistical properties of optical surfaces using a FECO scanning interferometer, *Applied Optics, Vol. 15*, pp. 2705–2721 (1976).

55. R.M. Pettigrew and F.J. Hancock, An optical profilometer, *Precision Engineering, Vol. 1*, p. 133 (1979).

56. M.J. Downs, W.H. McGivern and H.J. Ferguson, Optical system for measuring the profiles of super-smooth surfaces, *Precision Engineering, Vol. 7*, No. 4, pp. 211–215 (1985).

57. G.E. Sommargren, Optical heterodyne profilometry, *Applied Optics, Vol. 20*, pp. 610–618 (1981).

58. D.Pantzer, J. Politch and L. Ek, Heterodyne profiling instrument for the ångström region, *Applied Optic, Vol. 25*, pp. 4168–4172 (1986).

59. D. Pantzer, Step response and spatial resolution of an optical heterodyne profiling instrument, *Applied Optic, Vol. 26*, 1987, 3915–3918.

60. P. Hariharan, *Optical Interferometry*, Academic Press, Australia (1985).

61. J.H. Bruning, D.R. Herriott, J.E. Gallagher, D.P. Rosenfeld, A.D. White and D.J. Brangaccio, Digital wavefront measuring interferometer for testing optical surface and lenses, *Applied Optics, Vol. 13*, No. 11, pp. 2693–2703 (1974).

62. R.W. Peterson, G.M. Robinson, R.A. Carlsen, C.D. Englund, P.J. Moran and W.M. Wirth, Interferometeric measurement of the surface profile of moving samples, *Applied Optics, Vol. 23*, No. 10, pp. 1464–1466 (1984).

63. J.C. Wyant, C.L. Koliopoulos, B. Bhushan and O.E. George, An optical profilometer for surface characterisation of magnetic media, *ASLE Transactions, Vol. 27*, p. 101 (1984).

64. B. Bhushan, J.C. Wyant and C. L. Koliopoulos, Measurement of surface topography of magnetic tapes by Mirau interferometry, *Applied Optics, Vol. 24*, No. 10, pp. 1489–1497 (1985).

65. D.M. Perry, G.M. Robinson and R.W. Peterson, Measurement of surface topography of magnetic recording materials through computer analyzed

microscopic interferometry, *IEEE Transactions on Magnetics, Vol. Mag-19,* p. 1656 (1983).

66. D.M. Perry, P.J. Moran and G.M. Robinson, Three-dimensional surface metrology of magnetic recording materials through direct-phase detecting microscopic interferometry, *Journal of the Institution of Electronic and Radio Engineers, Vol. 55,* No. 4, pp. 145–150 (1985).

67. J.C. Wyant, C.L. Koliopoulos, B. Bhushan, D. Basila, Development of a three-dimentional noncontact digital optical profiler, *Journal of Tribology, Transactions of the ASME, Vol. 108,* No. 1, pp. 1–8 (1986).

68. S.R. Lange and B. Bhushan, Use of two- and three-dimensional noncontact surface profiler for tribology applications, *Surface Topography, Vol. 1,* No. 3, pp. 277–289 (1988).

69. J.C. Wyant and K. Creath, Advances in interferometric optical profiling, *International Journal of Machine Tools and Manufacture, Vol. 32,* No. 1/2, 5–10 (1992).

70. F. Francon, *Optical Interferometry.* Academic Press, New York (1966).

71. F. Francon and S. Mallick, *Polarization interferometers.* Wiley, New York (1971).

72. D.L. Lessor, J.S. Hartman and R.L. Gordon, Quantitative surface topography determination by Nomarski reflection microscopy. 1: Theory, *Journal of Optical Society of America, Vol. 69,* No. 2, pp. 357–366 (1979).

73. J.S. Hartman, R.L. Gordon and D. L. Lessor, Quantitative surface topography determination by Nomarski reflection microscopy. 2: Microscope modification, calibration, and planar sample experiments, *Applied optics, Vol. 19,* No. 17, pp. 2998–3009 (1980).

74. T.C. Bristow, Surface roughness measurements over long scan lengths, *Surface Topography, Vol. 1,* No.1, pp. 85–89 (1988).

75. G. Makosch and B. Drollinger, Surface profile measurement with a scanning differential a/c interferometer, *Applied Optics, Vol. 23,* No. 24, pp. 4544–4553 (1984).

76. Topo-3D – Non-contact microsurface measurement systems, (Product information), Wyko (1990).

77. Maxim.3D Model 57000 – Noncontact surface profile, (Product information), Zygo (1991).

78. Promap-512 – Optical profile, (Product information), Micromap (1991).

79. MP2000 – Non-contact surface profile, (Product information), Chapman (1991).

80. G. Binnig, H. Rohrer, C. Gerber and E. Weibel, Tunnelling through a controllable vacuum gap, *Applied Physics Letters, Vol. 40,* pp. 178–180 (1981).

81. G. Binnig H. Rohrer, C. Gerber and E. Weibel, Surface studies by scanning tunnelling microscopy, *Physical Review Letters, Vol. 49*, No. 1, pp. 57–61 (1982).

82. G. Binnig and H. Rohrer, Scanning tunnelling microscopy, *Surface Science Vol. 126*, pp. 236–244 (1983).

83. G. Binnig and C.F. Quate and C. Gerber, Atomic force microscope, *Physical Review Letters, Vol. 56*, No.9, pp. 930–933 (1986).

84. R.D. Young, Surface microtopography, *Physics Today, Vol. 24*, No. 11, pp. 42–49 (1971).

85. R.D. Young, Eight techniques for the optical measurement of surface roughness' pp. 73–219, U.S. Department of Commerce National Bureau of Standards (May 1973).

86. K. Mitsui, In-Process sensors for surface roughness and their applications, *Precision Engineering, Vol. 8*, No. 4, pp. 212–220 (1986).

87. T.V. Vorburger, Measurements of roughness of very smooth surfaces, *Annals of the CIRP, Vol. 36*, pp. 503–509 (1987).

88. D.J. Whitehouse, Instrumentation for measuring finish, defects and gloss, *Proceedings of SPIE, Vol. 525*, pp. 106–123 (1985).

89. J.C. Wyant, Optical profilers for surface roughness, *Proceedings of SPIE, Vol. 525*, pp. 174–181 (1985).

90. J.M. Bennett, L. Mattsson, *Introduction to Surface Roughness and Scattering.* Optical Society of America, Washington, D.C. (1989).

91. R.D. Young, T.V. Vorburger and E.C. Teague, In-process and on-line measurement of surface finish, *Annals of the CIRP, Vol. 29* pp. 435–440 (1980).

92. D.J. Whitehouse, Surface Metrology instrumentation, *Journal of Physics E: Scientific Instrument, Vol. 20*, pp. 1145–1155 (1987).

93. D.J. Whitehouse, Comparison between stylus and optical methods for measuring surfaces, *Annals of the CIRP, Vol. 37* pp. 649–653 (1988).

94. C.F. Quate, Vacuum tunnelling: a new technique for microscopy, *Physics Today*, pp. 26–33 (August 1986).

95. C. Schneiker, S. Hameroff, M. Voelker, J. He, E. Dereniak and R. McCuskey, Scanning tunnelling engineering, *Journal of Microscopy, Vol. 152*, Part 2, pp. 585–596 (1988).

96. H.K. Wickramasinghe, Scanning probe microscopy: current status and future trends, *Journal of Vacuum Science and Technology, Vol. A 8*, No. 1, pp. 363–368 (1990).

97. G.E. Sommargren, An optical measurement of surafce profile, *Precision Engineering, Vol. 3*, pp. 131–136 (1981).

98. E.C. Teague, T.V. Vorburger and D. Maystre, Light scattering from manufactured surfaces, *Annals of the CIRP, Vol. 30* pp. 563–569 (1981).

99. E.G. Thwaite, The extension of angular scattering techniques to the meaurement of intermediate scale roughness, *Annals of the CIRP, Vol. 31* pp. 463–465 (1982).

100. K.J. Stout, Optical assessment of surface roughness: The effectiveness of a low-cost, commercially available instrument, *Precision Engineering, Vol. 6*, No. 2, pp. 35–39 (1984).

101. C. Lukanowicz, Analysis of light scattering by rough manufactured surfaces, *Precision Engineering, Vol. 7*, No. 2, pp. 67–71 (1985).

102. R. Brodmann, Roughness form and waviness measurement by means of light scattering, *Precision Engineering, Vol. 8*, No. 4, pp. 221–226 (1986).

103. M. Shiraishi, A consideration of surface roughness measurement by optical method, *Journal of Engineering for Industry, Transactions of the ASME, Vol. 109*, pp. 100–105 (1987).

104. R.Brodmann, An optical instrument for measuring the surface roughness in production control, *Annals of the CIRP, Vol. 33* pp. 403–406 (1984).

105. K. Yanagi and M. Nakamura, An indirect optical measuring method of surface roughness on internal ground surface, *Bulletin of Japan Society of Precision Engineering, Vol. 21*, No. 3, pp. 217–218 (1987).

106. H. Tipton and J.I. Roberts, New optical method of assessing surface quality, *Proceedings of the Institution of Mechanical Engineers, Vol. 182*, Part 3K, pp. 274–278 (1967–8).

107. C. Gorecki, Optical classification of machined metal surfaces by fourier spectrum sampling, *Wear, Vol. 137*, pp. 287–298 (1990).

108. D.L. Jordan, R.C. Hollins, E. Jakeman and A. Prewett, Visible and infra-red scattering from well-characterised surfaces, *Surface Topography, Vol. 1*, No.1, pp. 27–36 (1988).

109. K.J. Stout, P.J. Sullivan, W.P. Dong, E. Mainsah, *Analysis of 3-D surface topography questionnaire, Interim Report of EC Contract No. 3374/1/0/170/90/2*, The University of Birmingham (September 1991).

110. K.J. Stout, E.J. Davis and P.J. Sullivan, *Atlas of Machined Surfaces*. Chapman and Hall (1990).

111. W.P. Dong, P.J. Sullivan and K.J. Stout, Comprehensive Study of Parameters for characterising 3-D Surface Topography – Part 1: Some Inherent Properties of Parameter Variation, Vol. 159, pp. 161–171 (1992).

112. Surface roughness – Terminology – Part 1: Surface and its parameters. *International Standard ISO 4287/1* (1984).

113. Method for the assessment of surface texture. *British Standard BS 1134* (1988).

114. Surface texture: surface roughness, waviness and lay. *American Standard ANSI B.46.1* (1978).

115. E.J. Davis, P.J. Sullivan and K.J. Stout, The application of 3-D topography to engine bore surfaces, *Surface Topography, Vol. 1*, No. 2, pp. 229–251 (1988).

116. R.S. Sayles and T.R. Thomas, Surface topography as nonstationary random process, *Nature, Vol. 271*, No.2, pp. 431–434 (1978).

117. T.R. Thomas and R.S. Sayles, Some problems in the tribology of rough surfaces, *Tribology International*, , pp. 163–168 (1978).

118. R.A. Agarwal, G.S. Patki and S.K. Basu, An analysis of surface profiles for stationarity and ergodicity, *Precision Engineering, Vol.1*, No. 3 (1979).

119. W.P. Dong, P.J. Sullivan and K.J. Stout, *Comprehensive Study of Parameters for Characterising 3-D Surface Topography – Part 2: Statistical Properties of Parameter Variation, Wear, Vol. 167*, pp. 9–21 (1993).

120. K.J. Stout and P.J. Sullivan, The analysis of the three-dimensional topography of the grinding process, *Annals of the CIRP, Vol. 39* pp. 545–548 (1989).

121. T. Tsukada and K. Sasajima, A three-dimensional measuring technique for surface asperities, *Wear, Vol. 71*, pp. 1–14 (1981).

122. B. Snaith, M.J. Edmonds and S.D. Probert, Use of a profilometer for surface mapping, *Precision Engineering, Vol. 3*, No. 2, pp. 87–90 (1981).

123. N. Idrus, An integrated digital system for three-dimensional surface scanning, *Precision Engineering, Vol. 3*, No. 1, pp. 37–43 (1981).

124. E.C. Teague, F.E. Scire, S. M. Baker and S.W. Jensen, Three-dimensional stylus profilometry, *Wear, Vol. 83*, pp. 1–12 (1982).

125. A.F. George and S.J. Radcliffe, Automated wear measurement on computerized profilometer, *Wear, Vol. 83*, pp. 327–337 (1982).

126. M. Chuard, A.C. Roudot and J. Mignot, On the use of a modular system for microtopographical surface measurement, *Wear, Vol. 96*, pp. 31–44 (1984).

127. A. Bengtsson and A. Ronnberg, Wide range three-dimensional roughness measuring system, *Precision Engineering, Vol. 6*, No. 3, pp. 141–147 (1984).

128. L.D. Chiffre and H.S. Nielsen, A digital system for surface roughness analysis of plane and cylindrical parts, *Precision Engineering, Vol. 9*, No. 2, pp. 59–64 (1987).

129. M. Chuard, J. Mignot, Ph. Nardin, D. Rondot, Range expansion and automation of a classical profilometer, *Journal of Manufacturing Systems, Vol. 6*, No. 3, pp. 223–231 (1987).

130. J.T. Hatazawa, K.Yamada and T. Kawaguchi, Three-dimensional observation and measurement of worn surfaces, *Surface Topography, Vol. 1*, No. 1, pp. 47–60 (1988).

131. S.J. Radcliffe and A.F. George, The analysis and presentation of multi-trace data, and some applicatins in industrial research, *Surface Topography, Vol. 1*, No. 2, pp. 215–227 (1988).

132. P.W. O'Callaghan, R.F. Babus'Haq, S.D. Probert and G.N. Evans, Three-dimensional surface topography assessments using a stylus/computer system, *International Journal of Computer Applications in Technology, Vol. 2*, No. 2, 101–107 (1989).

133. E.J. Davis, K.J. Stout and P.J. Sullivan, The scope of three-dimensional topography, *Industrial Metrology, Vol. 1*, pp. 193–216 (1990).

134. T. Kanada, T. Kubota and A. Suzuki, A three-dimensional surface profile measuring system with a specimen-levelling device, *Measurement Science Technology, Vol. 2*, pp. 191–197 (1991).

135. P.J. Sullivan, V. Poroshin and C. Hooke, Application of a three-dimensional surface analysis system to the prediction of asperity interaction in metallic contacts, *International Journal of Machine Tools and Manufacturing, Vol. 32*, No. 1/2, pp. 157–169 (1992).

136. The 3-D automated surface topography and analysis system (Product information), 3-D Digital Design and Development Ltd (1991).

137. Perthometer S8P – Surface measuring instrument for the acquisition, graphical presentation, evaluation and documentation of surface profiles. (Product information), Feinprüf Perthen (1991).

138. Surfascan 3D – Profilometer universel d'états de surface bi- et tridimensionnel (Product information), Somicronic (1991).

139. Surfcom 575-3D – Surface texture and contour measuring instrument (Product information). Tokyo Seimitsu (1992).

140. J.L. Guerrero, J.T. Black, Stylus resolution and surface damage as determined by scanning electron microscopy, *Journal of Engineering for Industry, Transactions of the ASME*, pp. 1087–1093 (November 1972).

141. D.J. Whitehouse, Some ultimate limits on the measurement of surface using stylus techniques, *Measurement and Control, Vol. 8*, pp. 147–151 (1975).

142. T.R. Thomas, Some examples of the versatility of stylus instruments, *Mécanique, Matériaux, Électricité, Vol. 337*, pp. 17–25 (1978).

143. J.I. McCool, Assessing the effect of stylus-tip radius and flight on surface topography measurements, *Journal of Engineering for Industry, Transactions of the ASME, Vol. 106*, pp. 106–110 (1984).

144. I. Sherrington and E.H. Smith, A quantitative study of the influence of stylus shape and load on the fidelity of data recorded by stylus instruments, *Proceedings of 2nd National Conference on Production Research, Edinburgh* (September 1986).

145. I. Elewa and M.M. Koura, Importance of checking the stylus radius in the measurement of surface roughness, *Wear, Vol. 109*, pp. 401–410 (1986).

146. D.E. Williamson and L.P. Mich, Tracer-point sharpness as affecting roughness measurements. *Transactions of the ASME*, pp. 319–323 (May 1947).

147. V. Radhakrishnan, Effect of stylus radius on the roughness values measured with tracing stylus instruments. *Wear, Vol. 16*, pp. 325–335 (1970).

148. Instruments for the measurement of surface roughness by the profile method – contact (stylus) instruments of progressive profile transformation – profile recording instruments. *International Standard ISO 1880* (1979).

149. Assessment of surface texture: methods and instrumentation. *British Standard BS 1134* (1988).

150. Leigh Mummery, *Surface Texture Analysis – The Handbook*. Hommelwerke GmbH (1990).

151. T.V. Vorburger and J. Raja, Surface finish metrology tutorial, *National Institute of Standards and Technology Internal Report (NISTIR 89/4088)* (1989).

152. E.J. Davis, Ph.D. Thesis, Coventry Polytechnic (1985).

153. P.J. Sullivan and N. Luo, The use of digital techniques for error correction in surface roughness measurement, *Surface Topography, Vol. 2*, No. 2, pp. 143–157 (1989).

154. I. Sherrington and E.H. Smith, Performance assessment of stylus-based areal roughness measurement systems, *International Journal of Machine Tools and Manufacture, Vol. 32*, No. 1/2, pp. 219–226 (1992).

155. S. Uchida, H. Sato and M.O. Hori, Two dimensional measurement of surface roughness by the light sectioning method, *Annals of the CIRP, Vol. 28* pp. 419–423 (1979).

156. H. Sato and M.O. Hori, Characteristics of two dimensional surface roughness – Taking self-excited chatter marks as objective, *Annals of the CIRP, Vol. 30*, pp. 481–486 (1981).

157. A.E. Ennos and M.S. Virdee, High-accuracy profile measurement of quasi-conical mirror surfaces by laser autocollimation, *Precision Engineering, Vol. 4*, No. 1, pp. 5–8 (1981).

158. M.S. Virdee, Nanometrology of optical flats by laser autocollimation, *Surface Topography, Vol.1*, No. 4, 1988, 415–425.

159. A. Boyde, Confocal optical microscopy, *Microscopy and Analysis*, pp. 7–13 (January 1988).

160. N.J. McCormick, Confocal scanning optical microscopy in materials science, *Microscopy and Analysis* , pp. 13–16 (November 1990).

161. D. Shotton, Confocal microscopy: The way ahead, *Microscopy and Analysis*, pp. 7–12 (July 1989).

162. L. Barsanti, P. Coltelli, V. Passarelli and P. Gualtieri, Applications in digital microscopy, *Microscopy and Analysis*, pp. 19–21 (November 1990).

163. M.J. Cookson, J.G. Holman, Computer based 3-D reconstruction for biomedical applications, *Image Enhancement & Analysis* pp. 15–17 (October 1990).

164. E.R. Weibel, Measuring through the microscope: development and evolution of stereological methods, *Journal of Microscopy, Vol. 155*, Part 3, pp. 393–403 (1989).

165. S.J. Jones and M.L. Taylor, Confocal fluorescence microscopy: some applications in bone cell biology, *Journal of Microscopy, Vol. 158*, Part 2, pp. 249–259 (1990).

166. X. Ronot and G. Raymond, Anchorage dependent laser cytometry for functional assessment of living cells, *Microscopy and Analysis*, pp, 41–43 (March 1991).

167. A.C. Scott, Geological application of laser scanning microscopy, *Microscopy and Analysis*, pp. 17–19 (March 1989).

168. V. Howard, Real 3-D measurements in microscopy using geometrical probes, *Microscopy and Analysis*, pp. 15–17 (November 1987).

169. H.T.M. van der Voort, G.J. Brakenhoff and M.W. Baarslag, Three-dimensional visualization methods for confocal microscopy, *Journal of Microscopy, Vol. 153*, Part 2, pp. 123–132 (1989).

170. F.S. Fay, W.Carrington and K.E. Fogarty, Three-dimensional molecular distribution in single cells analysed using the digital imaging microscope, *Journal of Microscopy, Vol. 153*, Part 2, pp. 133–149 (1989).

171. G.J. Brakenhoff, H.T.M. van der Voort, E.A. van Spronsen and N. Nanninga, Three-dimensional imaging in fluorescence by confocal scanning microscopy, *Journal of Microscopy, Vol. 153*, Part 2, pp. 151–159 (1989).

172. T. Wilson, Three-dimensional imaging in confocal systems, *Journal of Microscopy, Vol. 153*, Part 2, pp. 161–169 (1989).

173. K. Carlsson and A. Liljeborg, A confocal laser microscope scanner for digital recording of optical serial sections, *Journal of Microscopy, Vol. 153*, Part 2, pp. 171–180 (1989).

174. P.V. Oostveldt and S. Bauwens, Quantitative fluorescence in confocal microscopy, *Journal of Microscopy, Vol. 158*, Part 2, pp. 121–132 (1990).

175. R.W.W.V. Resandt, H.J.B. Marsman, R. Kaplan, J. Davoust, E.H.K. Stelzer and R. Stricker, Optical fluorescence microscopy in three dimensions: microtomoscopy, *Journal of Microscopy, Vol. 138*, Part 1, pp. 29–34 (1985).

176. M. Petran, M. Hadravsky, M.D. Egger and R. Galambos, Tandem-scanning reflected-light microscope, *Journal of the Optical Society of America, Vol. 58*, No. 5, pp. 661–664 (1968).

177. J.F. Nankivell, The theory of electron stereo-microscopy, *Optik, Vol. 20,* p. 172 (1963).

178. B. Hudson, The application of stereo-techniques to electron micrographs, *Journal of Microscopy, Vol. 98,* Part 3, pp. 396–401 (1973).

179. A. Boyde, Quantitative photogrammetric analysis and qualitative stereoscopic analysis of SEM images, *Journal of Microscopy, Vol. 98,* Part 3, pp. 452–471 (1973).

180. A. Boyde, P.G.T. Howell and F. Franc, Simple SEM stereophotogrammetric method for three-dimensional evaluation of features on flat substrates, *Journal of Microscopy, Vol 143,* Part 3, pp. 257–264 (1986).

181. J.A. Swift, Measuring surface variations with the scanning electron microscope using lines of evaporated metal, *Journal of Physics E: Scientific Instruments, Vol. 9,* pp. 804–804 (1976).

182. Y. Matsuno, H. Yamada, M. Harada and A. Kobayashi, The microtopography of the grinding wheel surface with SEM, *Annals of the CIRP, Vol. 24,* No. 1, pp. 237–242 (1975).

183. J.T.L. Thong and B.C. Breton, A topography measurement instrument based on the scanning electron microscope, *Review of Scientific Instrument, Vol. 63,* No. 1, pp. 131–138 (1992).

184. N.K. Myshkin, A.Ya. Grigoriev and O.V. Kholodilov, Quantitative analysis of surface topography using scanning electron microscopy, *Wear, Vol. 153,* pp. 119–133 (1992).

185. H. Sato and M. O-Hori, Surface roughness measurement by scanning electron microscope, *Annals of the CIRP, Vol. 1,* No.1, pp. 457–462 (1982).

186. H. Sato and M. O-Hori, Surface roughness measurement using scanning electron microscope with digital processing, *Journal of Engineering for Industry, Transactions of the ASME, Vol. 109,* pp. 106–110 (1987).

187. F. Rasigni, M. Varnier, J.P. Palmari and A. Llebaria, Spectral density function of the surface roughness for polished optical surfaces, *Journal of Optical Society of America, Vol 73,* No. 10, pp. 1235–1239 (1983).

188. I.C. Carlsen, Reconstruction of true surface topographies in scanning electron microscopes using back-scattered electrons, *Scanning, Vol. 7,* pp. 169–177 (1985).

189. J.Z. Raski, J.A. Levitt and K.C. Ludema, Surface topography from back-scattered electron data using a noise-reducing fourier method, *Surface Topography, Vol. 1,* No. 4, pp. 397–414 (1988).

190. Stereoscan 260/360 – Scanning electron microscope, (Product information), Leica Cambridge Ltd (1992).

191. SEMPROBE – An automated analytical scanning electron microscope, (Product information), Cameca Ltd, 1992.

192. SEM-IPS 10/500 – Image analysis systems for SEM, (Product information), Kontron Bildanalyse Ltd (1992).

193. A. Bartolome, R. Garcia, L. Vazquez and A.M. Baro, Imaging an optical disc by the combined use of scanning tunnelling microscopy and scanning electron microscopy, *Journal of Microscopy, Vol. 152*, Part 1, pp. 205–211 (1988).

194. R.A. Dragoset, R.D. Young, H.P. Layer, S.R. Mielczarek, E.C. Teague and R.J. Celotta, Scanning tunnelling microscopy applied to optical surfaces, *Optical Letters, Vol. 11*, No. 9, pp. 560–562 (1988).

195. M. Gehrtz, H. Strecker and H. Grimm, Scanning tunnelling microscopy of machined surfaces, *Journal of Vacuum Science and Technology, Vol. A6*, No. 2, pp. 432–435 (1988).

196. F. Besenbacher, E. Laegsgaard and I. Stensgaard, The scanning tunnelling microscope, *Microscopy and Analysis*, pp. 17–20 (July 1989).

197. S.P. Tear, The use of STM for surface structure determination, *Microscopy and Analysis*, pp. 7–9 (September 1990).

198. M. Hietschold, P.K. Hansma and A.L. Weisenhorn, Scanning-probe-microscopy and spectroscopy in materials science, *Microscopy and Analysis*, pp. 25–27 (September 1991).

199. J.K. Gimzewski, A. Humbert and D. W. Pohl, Scanning tunnelling microscopy of nanocrystalline silicon surfaces, *Surface Science, Vol. 168*, pp. 795–800 (1986).

200. M.J. Miles and H.W. Wills, The application of STM/AFM to biological molecules, *Microscopy and Analysis*, pp. 7–9 (July 1990).

201. R. Garcia, Scanning tunnelling microscopy in biology: changing the pace, *Microscopy and Analysis* pp. 27–29 (July 1991).

202. B.L. Blackfor, M.O. Watanabe, M.H. Jericho and D.C. Dahn, STM imaging of the complete bacterial cell sheath of Methanospirillu hungatei, *Journal of Microscopy, Vol. 152*, Part 1, pp. 237–243 (1988).

203. C.A.J. Putman, K.O.V.D. Werf, B.G.D. Grooth and N.F.V. Hulst, Atomic force microscope with integrated optical microscope for biological applications, *Review of Scientific Instrument, Vol. 63*, No. 3, pp. 1914–1917 (1992).

204. S. Coles, STM in the UK, *Microscopy and Analysis*, pp. 39–40 (July 1990).

205. O. Albrekstsen, L.L. Madsen, J. Mygind and K.A. Morch, A compact scanning tunnelling microscope with thermal compensation, *Journal of Physics E: Scientific Instrument, Vol. 22*, pp. 39–42 (1989).

206. S. Grafstrom, J. Kowalski and R. Neumann, Design and detailed analysis of a scanning tunnelling microscope, *Measurement Science and Technology, Vol. 1*, pp. 139–146 (1990).

207. S. Ito, T. Takenobu, H. Miyamoto, S. Mishima, H. Shimazu, T. Takase and T. Okada, Simultaneous observation on fine processing surface by an integrated scanning tunnelling microscope system with an optical microscope, *Journal of Japan Society of Precision Engineering, Vol. 25*, No. 2, pp. 156–157 (1991).

208. *The Scanning-Probe Microscope Book*. Burleigh Instruments, Inc. (1991).

209. I.H. Musselman, P.A. Peterson and P.E. Russell, Fabrication of tips with controlled geometry for scanning tunnelling microscopy, *Precision Engineering, Vol. 12*, No. 1, pp. 3–6 (1990).

210. D.K. Biegelsen, F.A. Ponce, J.C. Tramontana and S.M. Koch, Ion milled tips for scanning tunnelling microscopy, *Applied Physics Letters, Vol. 50*, No. 11, pp. 696–698 (1987).

211. T. Hashizume, I. Kamiya, Y. Hasegawa, N. Sano, T. Sakurai and H.W. Pickering, A role of a tip geometry on STM images, *Journal of Microscopy, Vol. 152*, Part 2, pp. 347–354 (1988).

212. NanoScope II – Scanning tunnelling microscope, (Product information), Digital Instruments, Inc. (1991).

213. Y. Kuk and P.J. Silverman, Scanning tunnelling microscope instrumentation, *Review of Scientific Instrument, Vol. 60*, No. 2, pp. 165–180 (1989).

214. J.E. Demuth, U.Koehler and R.J. Hamers, The STM learning curve and where it may take us, *Journal of Microscopy, Vol. 152*, Part 2, pp. 299–316 (1988).

215. M. Stedman, Limits of topographic measurement by the scanning tunnelling microscopes, *Journal of Microscopy, Vol. 152*, Part 3, pp. 611–618 (1988).

216. J.R. Matey and J. Blanc, Scanning capacitance microscopy, *Journal of Applied Physics, Vol. 57*, No. 5, pp. 1437–1444 (1985).

217. C.D. Bugg and P.J. King, Scanning capacitance microscopy, *Journal of Physics E: Scientific Instrument, Vol. 21*, pp. 147–151 (1988).

218. C.D. Bugg and P.J. King, Correcting scanning capacitance microscope images for the effect of surface gradient, *Precision Engineering, Vol. 12*, No. 4, pp. 239–244 (1990).

219. K.F. Sherwood and J.R. Crookall, Surface finish assessment by an electrical capacitance technique, Proceedings of the Institution of Mechanical Engineers, Vol. 182, Part 3K, pp. 344–349 (1967–8).

220. J.N. Brecker, R.E. Fromson and L.Y. Shum, A capacitance-based sruface texture measuring system, *Annals of the CIRP, Vol. 25*, pp. 375–377 (1977).

221. A.G. Lieberman, T.V. Vorburger, C.H.W. Giauque, D.G. Risko, and R. Resnick, Capacitance versus stylus measurements of surface roughness, *Surface Topography, Vol. 1*, No. 3, pp. 315–330 (1988).

222. J.L. Garbini, J.E. Jorgensen, R.A. Downs and S.P. Kow, Fringe-field capacitive profilometry, *Surface Topography, Vol. 1*, No. 1, pp. 99–110 (1988).

223. C.C. Williams and H.K. Wickramasinghe, Scanning thermal profile, *Applied Physics Letters, Vol. 49*, pp. 1587–1589 (1987).

224. K. Dransfeld and J.Xu, The heat transfer between a heated tip and a substrate: fast thermal microscopy, *Journal of Microscopy, Vol. 152*, Part 1, pp. 35–42 (1988).

225. R.A. Lemons and C.F. Quate, Acoustic microscopy, *Physical Acoustics, Vol. 14*, pp. 1–92 (1979).

226. H. Wickramasinghe, Scanning acoustic microscopy: a review, *Journal of Microscopy, Vol. 129*, pp. 63–73 (1983).

227. G.A.D. Briggs, P.J. Jenkins and M. Hoppe, How fine a surface crack can you see in a scanning acoustic microscope, *Journal of Microscopy, Vol. 159*, Part 1, pp. 15–32 (1990).

228. M.G. Somekh, Scanning Accoustic Microscopy, *Microscopy and Analysis*, pp. 7–10 (July 1988).

229. S.M. Pandit, F. Nassirpour and S.M. Wu, Stochastic geometry of anisotropic random surfaces with application to coated abrasives, *Journal of Engineering for Industry, Transactions of the ASME, Vol. 99B*, pp. 218–224 (1977).

230. Y.Y. Huang and S.M. Wu, Grinding surface characterisation by CEST, *International Journal of Machine Tool Design and Research, Vol. 26*, No. 4, pp. 431–444 (1986).

231. W.R. DeVries, A three-dimensional model of surface asperties developed using moment theory, *Journal of Engineering for Industry, Transactions of the ASME, Vol. 104*, pp. 343–348 (1982).

PART II
THREE-DIMENSIONAL SURFACE TOPOGRAPHY
– REVIEW OF PRESENT AND FUTURE TRENDS

1. E. J. Davis, K. J. Stout and P. J. Sullivan; The scope of three-dimensional surface topography, *Industrial Metrology, Vol 1*, pp. 193–216 (1990).

2. P. J. Sullivan, V. Poroshin & C. J. Hooke, An application of a three-dimensional surface analysis system to the prediction of asperity interaction in metallic contacts, *5th International Conference on the Metrology and Properties of Engineering Surfaces, Leicester.* Pergamon Press, Oxford (April 1991).

3. T.V. Vorburger, Measurements of roughness of very smooth surfaces, *Annals of the CIRP, Vol. 36* pp. 503–509 (1987).

4. I. Sherrington, *The Measurement and Characterisation of Surface Topography.* Ph.D. Thesis, Lancashire Polytechnic (1985).

5. I. Sherrington, Modern measurement techniques in surface metrology. Part 1: Stylus instruments, electron microscopy and non-optical comparators, *Wear, Vol. 125*, pp. 271–288 (1988).

6. I. Sherrington, Modern measurement techniques in surface metrology. Part 2: Optical instruments, *Wear, Vol 125*, pp. 289–308 (1988).

7. K.J. Stout, P.J. Sullivan, W.P. Dong, E. Mainsah, K. Subari, and T. Mathia; The development of an integrated approach to 3-D surface finish assessment, *Interim Report No. 1, EC Contract No. 3374/1/0/170/90/2* (March 1991).

8. The University of Birmingham, Centre for Metrology, The development of an integrated approach to 3-D surface finish assessment, *Project proposal submitted to the EC by the University of Birmingham and the École Centrale de Lyon* (1989).

9. D. J. Whitehouse, Parameter rash – is there a cure?, *Wear, 83*, pp. 75–78 (1982).

10. A.G. Lieberman, T.V. Vorburger, C.H.W. Giauque, D.G.Risko, R. Resnick and J. Rose, Capacitance versus stylus measurements of surface roughness, *Surface Topography, Vol. 1*, No. 3, pp. 315–330 (1988).

11. J.L. Garbini, J.E. Jorgensen, R.A. Downs and S.P. Kow, Fringe-field capacitive profilometry, *Surface Topography, Vol. 1*, No. 1, pp. 99–110 (1988).

12. R. W. Wooley; Pneumatic probe – developed for precision surface measurement, *Precision Engineering, Vol. 12*, No. 3 (1990).

13. G.V. Blessing and D.G. Eitzen, Surface roughness sensed by ultrasound, *Surface Topography, Vol. 1*, No. 2, pp. 253–267 (1988).

14. N. Luo, *Fidelity Within Surface Measurement*. M.Phil. Thesis, University of Birmingham (1991).

15. T. Kanada, T. Kubota and A. Suzuki, A three-dimensional surface profile measuring system with a specimen-levelling device, *Measurement Science and Technology, Vol. 2*, pp. 191–197 (1991).

16. Y.Z. Hu, K. Tonder, Simulation of 3-D random rough surface by 2-D digital filter and Fourier analysis, *5th International Conference on the Metrology and Properties of Engineering Surfaces (BCR Report)*. EC Bureau of Reference, Brussels (April 1991).

17. I. Sherrington and E.H. Smith, Fourier models of the surface topography of engineering components, *Surface Topography, Vol. 1*, No. 1, pp. 11–25 (1988)

18. I. Sherrington and E.H. Smith, Areal Fourier analysis of surface topography. Part 1: Computational methods and sampling considerations, *Surface Topography, Vol. 3*, No. 1, pp. 43–68 (1990).

19. Surface topography users' questionnaire (private communication), Centre for Metrology, University of Birmingham, (May 1991).

20. J. G. Proakis, D. G. Manolakis. *Introduction to Digital Signal Processing*. Macmillan, New York (1988).

21. I. Sherrington and E.H. Smith. Parameters for characterising the surface topography of engineering components, *Proceedings of the Institution of Mechanical Engineers, Vol 201*, No. C4.

22. T.R. Thomas, *Rough Surfaces*. Longman (1982).

23. D.J. Whitehouse and J.F. Archard, The properties of random surfaces in contact, *Proceedings of the ASME Annual Winter Meeting*, pp. 16–20, (November 1969).

24. P.R. Nayak, Random process model of rough surfaces, *Journal of Lubrication Technology*, pp. 398–407, July (1971).

25. B. Nowicki, Multiparameter representation of surface roughness, *Wear, Vol. 102*, pp. 161–176 (1985).

26. W.P. Dong, P.J. Sullivan and K.J. Stout, Comprehensive study of parameters for 3-D surface topography analysis. Part 1: Inherent properties of parameter variation, *Wear, Vol. 159*, pp. 161–171 (1992).

27. W. Eatson and A. Woods, The 3-D representation of engineering surfaces, *Surface Topography, Vol. 1*, No. 2, pp. 165–182 (1988).

28. Bao Qian and Yuan Chang Liang, The application of modern time series analysis method in researching and measuring surface roughness, *Proceedings of the 2nd IMEKO*, pp. 173–177 (May 1989).

29. M. Santochi and M. Vignale, A study on the functional properties of a honed surface, *Annals of the CIRP, Vol. 31*, pp. 431–434 (1982).

30. M. Santochi and G. Tantussi, Surface parametrical microgeometry and functional models: a new approach, *Precision Engineering, Vol. 6*, No. 4, pp. 201–206 (1984).

31. E.J. Davis, P.J. Sullivan and K.J. Stout, The application of 3-D topography to engine bore surfaces, *Surface Topography, Vol. 1*, No. 2, pp. 229–251 (1988).

32. B.D. Boudreau and J. Raja, Analysis of lay characteristics of three-dimensional surface maps, *5th International Conference on Metrology and Properties of Engineering Surfaces, Leicester.* Pergamon Press, Oxford (1991).

33. K.J. Stout and P.J. Sullivan, The analysis of the three-dimensional topography of the grinding process, *Annals of the CIRP, Vol. 38*, pp. 545–548 (1989).

34. B.B. Mandelbrot, *Fractal Geometry of Nature.* Freeman, San Francisco (1977).

35. T.R. Thomas and A.P. Thomas, Fractals and engineering surface roughness, *Surface Topography, Vol. 1*, No. 2, pp. 143–152 (1988).

36. C.A. Brown and G. Savary, Describing ground surface texture using contact profilometry and fractal analysis, *Wear, Vol. 141*, pp. 211–226 (1991).

37. A. Majumdar and B. Bhushan, Fractal model of elastic-plastic contact between rough surfaces, *Journal of Tribology, Transactions of the ASME, Vol. 113*, pp. 1–11 (1991).

38. B.J. Griffiths, Manufacturing surface design and monitoring for performance, *Surface Topography, Vol. 1*, No. 1, pp. 61–69 (1988).

PART III
AN INTRODUCTION TO VISUALISATION TECHNIQUES AND A PRIMARY PARAMETER SET FOR CHARACTERISING THREE-DIMENSIONAL SURFACE TOPOGRAPHY

1. K.J. Stout, P.J. Sullivan, W.P. Dong, E. Mainsah, K. Subari and T. Mathia, The development of methods for the characterisation of roughness in 3 dimensions, *Interim Report No. 1, EC Contract No. 3374/1/0/170/90/2 (March 1991), First EC workshop on 3-D surface topography measurement and characterisation* (September 1991).

2. K.J. Stout, P.J. Sullivan, W.P. Dong, E.Mainsah, N. Luo, T. Mathia and H. Zahouani, *The Development of Methods for the Characterisation of Roughness in 3 Dimensions* Commission of the European Communities (BCR-3374/1/0/170/90/2) (1993).

3. M.S. Longuet-Higgins, The statistical analysis of a random, moving surface, *Philosophical Transactions of the Royal Society, Vol. 249*, Series A, pp. 321–384 (1957).

4. M.S. Longuet-Higgins, Statistical properties of an isotropic random surface, *Philosophical Transactions of the Royal Society, Vol. 250*, Series A, pp. 157–174 (1957).

5. P.R. Nayak, Random process model of rough surfaces, *Journal of Lubrication Technology Transactions of the ASME*, pp. 384–407 (July 1971).

6. J. Peklenik and M. Kubo, A basic study of a three-dimensional assessment of the surface generated in a manufacturing process, *Annals of the CIRP, Vol. 16*, pp. 257–265 (1968).

7. M. Kubo and Peklenik, An analysis of micro-geometrical isotropy for random surface structures, *Annals of the CIRP, Vol. 16*, pp. 235–242 (1968).

8. J. Pekenik, New developments in surface characterization and measurements by means of random process analysis, *Proceedings of the Institution of Mechanical Engineers, Vol. 182*, Part 3K, pp. 108–126 (1967–8).

9. S.M. Pandit, F. Nassirpour and S.M. Wu, Stochastic geometry of anisotropic random surfaces with application to coated abrasives, *Journal of Engineering for Industry, Transactions of the ASME, Vol. 99B*, pp. 218–224 (1977).

10. R.S. Sayles and T.R. Thomas, The spatial representation of surface roughness by means of the structure function: A practical alternative to correlation, *Wear, Vol. 42*, pp. 263–276 (1977).

11. D.J. Whitehouse and M.J. Phillips, Discrete properties of random surfaces, *Philosophical Transactions of the Royal Society, Vol. 290*, Series A, pp. 267–298 (1978).

12. D. Wehbi, M.A. Clerc and C. Roques-Carmes, Three-dimensional quantification of wear tracks on amorphous NiB coatings' *Wear, Vol. 107*, No. 2, pp. 263–278 (1986).

13. E.J. Davis, P.J. Sullivan and K.J. Stout, The application of 3-D topography to engine bore surfaces, *Surface Topography, Vol. 1*, No. 2, pp. 229–251 (1988).

14. I. Sherrington and E.H. Smith, Fourier models of the surface topography of engineering components, *Surface Topogrpahy, Vol. 1*, No. 1, pp. 11–25 (1988).

15. B.D. Boudreau and J. Raja, Analysis of lay characteristics of three-dimensional surface maps, *Proceedings of the 5th International Conference on Metrology and Properties of Engineering Surfaces, Leicester,* Pergamon Press, Oxford pp. 171–177. (1991).

16. A. Bengtsson, *Three-Dimensional Measurement and Analysis of Surface Roughness of Curved Surfaces.* Chalmers University of Technology, July (1991).

17. S.R. Lange and B. Bhushan, Use of two- and three-dimensional noncontact surface profiler for tribology applications, *Surface Topography, Vol. 1*, No. 3, pp. 277–289 (1988).

18. W.P. Dong, P.J. Sullivan and K.J. Stout, The significance of surface features in characterisation 2-D and 3-D surface topography, *Engineered Surfaces, ASME, PED-Vol. 62*, pp. 1–15 (1992).

19. W.P. Dong, N.L. Luo, P.J. Sullivan and K.J. Stout, A proposal of parameters for characterising three-dimensional surface topography, *Research report distributed in academic and industry in Europe,* University of Birmingham (October 1992).

20. K.J. Stout, P.J. Sullivan, Workshop on the characterisation of surfaces in 3-D, *First EC workshop on 3-D surface topography measurement and characterisation* (September 1991).

21. BSI, Assessment of Surface Texture, Part 1. Methods and Instrumentation. *British Standard BS 1134* (1988).

22. ANSI, Surface Texture: Surface Roughness, Waviness and Lay. *American Standard ANSI B.46.1* (1985).

23. ISO, Surface Roughness – Terminology – Part 1: Surface and its Parameters. *International Standard ISO 4287/1* (1984).

24. T.R. Thomas, *Rough Surfaces.* Longman, London (1982).

25. K. Peucker, D. Douglas, Detection of surface-specific points by local parallel processing of discrete terain elevation data, *Computer Graphics and Image Processing, Vol. 4*, pp. 375–387 (1975).

26. P. J. Scott, A discussion on the characterizaton of areal measurements, *VIII International Colloqium, Chemnitz, Germany* (1992).

27. R.S. Sayles and T.R. Thomas, Measurements of the statistical microgeometry of engineering surfaces, *Journal of Lubrication Technology, Transactions of the ASME, Vol. 101*, pp. 409–418 (1979).

28. DIN, Measurement of Surface Roughness; Parameters Rk, Rpk, Rvk, Mr1, Mr2 for the Description of the Material Portion in the Roughness. *German Standard, DIN 4776* (1990).

29. K.J. Stout and P.J. Sullivan, The analysis of the three-dimensional topography of the grinding process, *Annals if the CIRP, Vol. 38*, pp. 545–548 (1989).

30. K.J. Stout, E.J. Davis and P.J. Sullivan, *Atlas of Machined Surfaces*. Chapman and Hall, London (1990).

PART IV
APPLICATIONS OF THREE-DIMENSIONAL
SURFACE METROLOGY

1. E.J. Davis, P.J. Sullivan and K.J. Stout, The application of 3-D topography to engine bore surfaces, *Surface Topography, Vol. 1*, No. 2, pp. 229–251 (1988).

2. K.J. Stout and P.J. Sullivan, The analysis of the 3-dimensional topography of a grinding process, *Annals of the CIRP, Vol. 38*, pp. 545–548 (1989).

3. J. Kagami, J.T. Hatazawa, K. Yamada and T. Kaeaguchi, Three-dimensional observation and measurement of worn surfaces, *Surface Topography, Vol. 1*, No. 1, pp. 47–60 (1988).

4. D. Wehbi, M.A. Clerc and C. Roques-Carmes, Three-dimensional quantification of wear tracks on amorphous NiB coatings, *Wear, Vol. 107*, No. 2, pp. 263–278 (1986).

5. D.K. Aspinwall, M.L.H. Wise, K.J. Stout, T.H.A. Goh, F.L. Zhao and M.F. El-Menshawy, Electrical discharge texturing, *International Journal of Machine Tools and Manufacture, Vol. 32*, No. 1/2, pp. 183–193 (1992).

6. G. Zhou, M. Leu, E. Geskin, Y. Chung and J. Chao, Investigation of topography of waterjet-generated surfaces, *Engineered Surfaces, ASME, PED-Vol. 62*, pp. 191–202 (1992).

7. P.J. Sullivan and L. Blunt, Three-dimensional characterisation of indentation topography – visual characterisation, *Wear, Vol. 159* (1992).

8. L. Blunt and P.J. Sullivan, The measurement of the topography of hardness indentations, *Tribology International* (1994).

9. L.E.C. van de Leemput *et al.*, Topography of YBa2Cu3O7-d single crystals, spectroscopy of thin films and sintered YBa2Cu3O7-d: theory and STM observations, *Journal of Microscopy, Vol. 152*, Part 1, pp. 103–115 (1988).

10. M.M. Dovek *et al.*, Observation and manipulation of polymers by scanning tunnelling and atomic force microscopy, *Journal of Microscopy, Vol. 152*, Part 1, 1988, Journal of Microscopy, Vol. 152, Part 1, 1988, 229–236.

11. J. Wintterlin, J. Wiechers, T. Gritsch, H. Hofer and R.J. Behm, Imaging of individual atoms on an Al(111) surface by scanning tunnelling microscopy, *Journal of Microscopy, Vol. 152*, Part 2, pp. 423–425 (1988).

12. M. Amrein, A. Stasiak, E. Gross and G. Travaglini, Scanning tunnelling microscopy of reA-DNA complexes coated with a conducting film, *Science, Vol. 240*, pp. 514–516 (1988).

13. J.G. Mantovani et al, Scanning tunnelling microscopy of tobacco mosaic virus on evaporated and sputter-coated palladium/gold substrates, *Journal of Microscopy, Vol. 158*, Part 1, pp. 109–116 (1990).

14. Perthometer S8P – Surface measuring instrument for the acquisition, graphical presentation, evaluation and documentation of surface profiles (Product information), Feinprüf Perthen (1991).

15. Surfascan 3D – Profilometer universel d'états de surface bi- et tridimensionnel (Product information), Somicronic (1991).

16. Surfcom 575-3D – Surface texture and contour measuring instrument (Product information), Tokyo Seimitsu (1992).

17. RM600 – A new approach to surface measurement (Product information), Optische Werke G. Rodenstock (1991).

18. Focodyn – Optical probe for perthometer non-contact profile acquisition (Product information), Feinprüf Perthen GmbH (1991).

19. UB16 – Precision optical length measurement system, (Product information), Ulrich Breitmeier Messtechnik GmbH (1991).

20. Topo-3D – Non-contact microsurface measurement systems, (Product information), Wyko (1990).

21. Maxim.3D Model 57000 – Noncontact surface profile, (Product information), Zygo (1991).

22. MP2000 – Non-contact surface profile, (Product information), Chapman (1991).

23. G. Binnig, H. Rohrer, C. Gerber and E. Weibel, Tunnelling through a controllable vacuum gap, *Applied Physics Letters, Vol. 40*, pp. 178–180 (1981).

24. NanoScope II – Scanning tunnelling microscope, (Product information), Digital Instruments, Inc. (1991).

25. *The Scanning-Probe Microscope Book. Burleigh Instruments, Inc. (1991).*

26. G. Binnig and C.F. Quate and C. Gerber, Atomic force microscope, *Physical Review Letters, Vol. 56*, No.9, pp. 930–933 (1986).

27. W.P. Dong, E. Mainsah, P.J. Sullivan and K.J. Stout, Instruments and measurement techniques of 3-dimensional surface topography, (Part I of this Monograph).

28. J.B.P. Williamson, Microtopography of surfaces, *Proceedings of the Institution of Mechanical Engineers, Vol. 182*, Part 3K, pp. 21–30 (1967–8).

29. J. Peklenik and M. Kubo, A basic study of a three-dimensional assessment of the surface generated in a manufacturing process, *Annals of the CIRP*, Vol. 16, pp. 257–265 (1968).

30. M. Minsky, Microscopy apparatus, United States Patent Office. Filed November 7 1957, granted December, 19 1961. Patent No. 3013467.

31. M.O. Dupuy, High-precision optical profilometer for the study of micro-geometrical surface defects, *Proceedings of the Institution of Mechanical Engineers*, Vol. 182, Part 3K, pp. 255–259 (1967–8).

32. W.P. Dong, N. Luo, P.J. Sullivan and K.J. Stout, A proposal of parameters for characterising three-dimensional surface topography, *Research report in circulation in European academia and industry* (1992).

33. G. Marsden, L Blunt and H Lewis, *Topography of Skin Tumours* (1994).

34. K.J. Stout and L. Blunt, *2nd International Conference on Surface Engineering, Nanometers to Microns; Surface Measurement in Bio-Engineering*, Adelaide, Australia (1994).

35. B.C. Wadell *Transmission Line Handbook 1991.* ARTEC House Inc (1991).

36. L Blunt, D, Holland and M.E.Yakinci Microstructure Development in Bulk and Thick Film Glass-Ceramic Superconductors' *Proceedings of the International Conference on IMIC,* Garmisch-Partenkirchen, Germany (1990).

37. J.C. Wyant, C.L. Koliopoulos, B. Bhushan and O.E. George, An optical profilometer for surface characterisation of magnetic media, *ASLE Transactions, Vol. 27*, p. 101 (1984).

General Index

2D analogue system 14
    surface topography, and 3D 6
    and 3D, analysis 61
3D digital system 14
    with two translation stages 15
3D measurement systems, history of 5
3D profile measurement, mechanisms of
    10 *et seq*
3D surface topography, advantage of 7-10
    areas of application 6
    disadvantages 10
    trends 67 *et seq*
3D surface topography instruments 4
    and computers 9
    measurement range 58
    measurement speed 60
    resolution 58
    spatial range 18
    stylus problems 60
    systems, features of 74
    vertical range 18

amplitude and height distribution
    parameters 104
arithmetic mean summit curvature surface
    115
astigmatic method 24
atomic force microscope (AFM) 5, 41,
    50-53
    and STM 53

bearings and fluid retention properties,
    functional parameters 118 *et seq*

characterisation
    functional 80
    gear surface 129
    hip prostheses surface 147
characterisation reference datum plane 78
characterisation techniques 90
characterisational parameters 78
charge injection device (CID) 36
clipping 100
closed loop position trigger 16

confocal laser scanning microscope 20
confocal method 28
contour plot 95
    of shaped surface 96
core fluid retention index 119
critical angle method 23

data acquisition 15-16
data logging 77
data sampling 76
datum plane definition 76
decurved topography 131
depth of focus method 28
developed interfacial area ratio 116
differential detection 23
differential interference contrast image
    (DIC) 37
digital filtering 77
digitisation 73
digitised surface, coordination of 91
dynamic measurement 12

EC
    Programme of Applied Metrology and
        Chemical Analysis xviii, xx
    survey 67 *et seq*
    workshop 90
electro-discharge machining (EDM) 9
electron microscopy 41 *et seq*
    properties 46
engine bore surface measurement 133
    decurved topography 136
    instrument 134
    topography 134

fastest decay autocorrelation length 114
focus detection 20
    disadvantages of 31
    instruments 5
    methods 30
    system 72
Foucalt method 24
fractal characterisation 83
functional characterisation 80

gear surface characterisation 129
    measurement 125, 129
gear tooth surface, surface height
    distribution 132
greyscale image 97

hip prostheses, measurement 145
    characterisation of surfaces 147
    instrument 147
human skin, measurement of 143
    topography of 144
hybrid parameter 115 *et seq*

indentation, intensity plot 155
    inverted isometric plot 157
    numerical characteristic 160
    surface topography characteristic 153
    visual characteristic 154
intensity detection 21
interference microscope 5
interferometers 32 *et seq*
    Fizea 34
    Linnick 34
    Michelson 33, 34
    phase shifting 34
    properties of 39
    scanning differential 37
    Wyko Topo 71
    Zygo Zerodur 71
interferometry 7
inversion 98
isometric plots 93
isotropy 108

kurtosis 106

levelling of specimen surface 71
    software 71
light scattering instruments 71
long crestedness 108

manipulation techniques 98 *et seq*
measurement
    dynamic (on the fly) 76
    speed 126
    static 76
Michelson interferometer 33
microscope, atomic force (AFM) 5, 41,
    50-53
    confocal laser scanning 20
    electron 41 *et seq*
    interference 5
    scanning acoustic (SAM) 41
    scanning capacitance (SCM) 41, 54-7
    scanning electron (SEM) 4, 41, 72
    scanning field emission 41
    scanning, non-optical 41 *et seq*

scanning probe 41, 46
scanning thermal 41
scanning transmission electron
    (STEM) 41
scanning tunnelling (STM) 5, 41, 47,
    48, 72
tandem scanning 30
transmission electron (TEM) 4, 41

Nankivell algorithm 43
non-parametric measurement 4

on-the-fly measurement 12, 76
open loop position trigger 16
open loop timing trigger 16
optical specular reflectance 4
optical systems 70
optical vs stylus instruments 126

parallax equation 44
parameter rash 83
parameter specification 103
parameter, hybrid 115 *et seq*
    set, primary 103
    spatial 106
    surface roughness 133
parameter set, primary 103
parametric measurement 4
phase shifting interferometer 34
polished brass surface
    measurement of 149
    topography of 151
primary parameter set 90
profiling 4

quantitative measurements 5

radial scan 12
raster scan 11
root mean square deviation 104
root mean square slope 115
root mean square (RMS) measurement 4

sampling data points, number 18
scan, radial 12
    raster 11
scanning acoustic microscope (SAM) 41
scanning capacitance microscope (SCM)
    41, 54-7
scanning differential interferometric
    instruments 37

scanning electron microscope (SEM) 4, 41, 72
scanning field emission microscope 41
scanning microscopy, non-optical 41 *et seq*
scanning probe microscope 41, 46
scanning thermal microscope 41
scanning transmission electron microscope (STEM) 5, 41
scanning tunnelling microscope (STM) 5, 41, 47, 48, 72
   atomic tip 49
SEM images, quantisation steropair technique 43
   direct integration 45
skew beam method 26
skewness 106
spatial parameters 106
specimen relocation 71
spectral analysis 79
spectral density function 71
static (point to point) measurement 13
statistical characterisation 79
steropair technique 43
stylus instruments 10 *et seq*, 17
   3D 17 *et seq*
   applications 70
   coordinate error 18
   dynamic characteristics 17
   problems with 6, 10
   skid, effect of 17
   resolution relation 75
summits, density of 107
surface bearing index 133
surface bearing ratio 118
   index 119
surface, functional performance and engineering application 82
surface image enhancement 101, 102
surface measurement, history of 4
surface replication 144
surface roughness parameters 133
surface topography,
   characteristics 3
   computers, use of 9
   evolution 3
   scope 69

tandem scanning microscope 30
ten point height 104
texture aspect ratio 108
texture direction 113

thick film superconductor, measurement of 139
time series analysis 79
tomography 3
topografiner 5, 41
topographical analysis, reference datum for 91
topography measuring instruments, range of 125
transmission electron microscope (TEM) 4, 41
trigger, closed loop position 16
   open loop timing 16
truncation 98
   of EDM surface 99

University of Birmingham Centre for Metrology 125, 127

valley fluid retention index 119
visual inspection 83
visualisation technique 92
   plots 93
zooming 100

Author Index

This author index contains

(i) page references referring to page numbers in the body of the text. These are characterised either as small roman numerals or ordinary page numbers eg vi or 136

(ii) bibliography references which contain the bibliography Part number followed by the reference number within that particular Part bibliography eg 2.111 refers to reference 111 in the bibliography for Part II.

Albrekstan, O. **1.205**
Amrein, M. **4.12**
Archard, J F. **2.23**
Aspinwall, D K. **4.5**

Baarslag, M W. **1.169**
Babus' Haq, R F. **1.132**
Baker, S M. **1.124**
Baro, A M. **1.193**
Barsanti, L. **1.162**
Bartolome, A. **1.193**
Basila, D. **1.67**
Basu, S K. **1.118**
Bauwens, S. **1.174**
Behm, R J. **4.11**
Bengtsson, A. **1.127**, **3.16**
Bennett, J A. **1.8**
Bennett, J M. **1.54**, **1.90**
Besenbacher, F. **1.196**
Bhashan, B. **1.64**, **1.67**, **1.68**, **2.37**, **3.17**, **4.37**
Biegelsen, D K. **1.210**
Binnig, G. 5, 47, 50, 126, 150, **1.80-82**, **1.83**, **4.23**, **4.26**
Black, J T. **1.140**
Blackfor, B L. **1.202**
Blanc, J. 55, **1.216**
Blessing, G V. **2.13**
Blunt, L. 153, **4.7**, **4.8**, **4.33**, **4.34**, **4.36**
Boudreau, B D. **2.32**, **3.15**
Bouwhuis, G. **1.43**

Boyde, A. **1.30**, **1.30**, **1.33**, **1.159**, **1.179**, **1.180**
Braat, J J M. **1.43**
Brakenhoff, G J. **1.169**
Brangaccio, D J. **1.61**
Brecker, J N. 55, **1.220**
Breton, B C. **1.183**
Briggs, G A D. **1.227**
Bristow, T C. 37, **1.74**
Brodmann, R. **1.102**
Brown, C A. **2.30**
Bruning, J H. 34, **1.61**
Bugg, C D. 56, 57, **1.127**, **1.128**

Carlsen, R A. **1.62**, **1.188**
Carlsson, K. **1.173**
Celotta, R J. **1.194**
Chao, J. **4.6**
Chiffre, C D. **1.128**
Chinmayanandam, T K. **1.9**
Chuard, M. **1.126**
Chung, Y. **4.6**
Clerc, M A. **3.12**, **4.4**
Coles, S. **1.204**
Coltelli, P. **1.162**
Cookson, M J. **1.163**
Cosslett, V E. **1.10**
Creath, K. **1.69**
Crookall, J R. 54, **1.219**

Dagwall, H. **1.2**
Dahn, D C. **1.202**
Davis, E J. **1.15**, **1.110**, **1.133**, **1.152**, **2.1**, **2.31**, **3.13**, **3.20**, **4.1**
Davoust, J. **1.175**
Demuth, J E **1.214**
Dereniak, E. **1.95**
De Vries, W R. **1.231**
Dinnis, A R. **1.32**
Dobosz, M. **1.46**, **1.47**
Dong, W. xv, **1.109**, **1.111**, **1.119**, **2.7**, **2.263**, **3.1**, **3.2**, **3.18**, **3.19**, **4.27**, **4.32**
Douglas, D. **3.25**
Dovek, M M. **4.10**
Downs, M J. **1.56**, **2.11**
Downs, R A. **1.222**
Dragoset, R A. **1.194**
Dransfield, K. **1.224**
Drollinger, B. **1.75**
Dupuy, M O. 5, 19, 126, **1.40**, **4.31**

Eatson, W. 2.27
Edmonds, M J. 1.122
Egger, M D. 1.176
Eitzen, D G. 2.13
Ek, L. 1.58
Elewa, I. 1.145
El-Menshawy, M F. 4.5
Englund, C D. 1.62
Ennos, A E. 1.157
Evans, G N. 1.132

Fainman, Y. 1.45
Fay, F S. 1.170
Ferguson, H J. 1.56
Franc, F. 1.180
Francon, M. 1.6, 1.70, 1.71
Fromson, R E. 1.220

Galambos, R. 1.177
Gallagher, J E. 1.01
Garbini, J L. 55, 1.222, 2.11
Garcia, R. 1.193, 1.201
Gehrtz, M. 1.195
George, A F. 1.125, 1.131, 4.37
Gerber, C. 1.81, 1.83, 4.26
Geskin, E. 4.6
Gimzewski, J K. 1.199
Giauque, C H W. 1.221, 2.10
Goh, T H A. 4.5
Gordon, R L. 1.72, 1.73
Gorecki, C. 1.51, 1.107
Grafstrom, S. 1.206
Griffiths, B J. 2.38
Grigoriev, A J. 1.184
Grimm, H. 1.195
Gritsch, T. 4.11
Grooth, B G D. 1.203
Gross, E. 4.12
Gualtieri, P. 1.162
Guerrero, J L. 1.140

Hadravsky, M. 30, 1.176
Haine, M E. 1.11
Hale, A J. 1.20
Hameroff, S. 1.95
Hamers, R J. 1.214
Hamilton, D K. 1.50
Hancock, F J. 1.55
Hansma, P K. 1.198
Harada, M. 1.182
Hariharan, P. 1.60

Hartman, J S. 1.72, 1.73
Hasegawa, Y. 1.211
Hashizume, T. 1.211
Hatazawa, J. 1.130
Hatazawa, T. 4.3
He, J. 1.95
Heidenreich, R D. 1.12
Hietschold, M. 1.198
Herriot, D R. 1.01
Hirsch, P B. 1.14
Hofer, H. 4.11
Holland, D. 4.36
Hollins, R C. 1.108
Holman, J G. 1.163
Hooke, C. 1.135, 2.2
Hoppe, M. 1.227
Hori, M O. 1.155, 1.156
Howard, V. 1.168
Howell, P G T. 1.33, 1.180
Hu, Y Z. 2.16
Huang, Y Y. 1.230
Hudson, B. 44, 1.178
Hulst, N F V. 1.203
Humbert, A. 1.199

Idrus, N. 1.123
Ito, S. 1.207

Jakeman, E. 1.108
Jenkins, P J. 1.227
Jensen, S W. 1.124
Jericho, M H. 1.202
Johari, O. 1.34
Jones, S J. 1.165
Jordan, D L. 1.108
Jorgenson, J E. 1.222, 2.11

Kaeaguchi, T. 4.3
Kagami, J. 4.3
Kamiya, I. 1.211
Kanada, T. 1.134, 2.15
Kaplan, R. 1.175
Kawaguchi, T. 1.130
Kholodilov, O V. 1.184
King, P J. 56, 1.217, 1.218
Knoll, M. 41, 1.15
Kobayashi, A. 1.182
Koch, S M. 1.210
Koehler, U. 1.214
Kohno, T. 1.49, 1.52
Kolipoulos, C L. 1.63, 1.64, 1.67, 4.37

Koura, M M. **1.145**
Kow, S P. **1.222, 2.11**
Kowalski, J. **1.206**
Kubo, M. 12, **1.25, 3.6, 3.7, 4.29**
Kubota, T. **1.134, 2.15**
Kuk, Y. **1.213**

Laegsgaard, E. **1.196**
Lange, S R. **1.68, 3.17**
Layer, H P. **1.194**
Lemons, R A. **1.225**
Lenz, E. **1.45**
Lessor, D L. 37, 38, **1.72, 1.73**
Leu, M. **4.6**
Levitt, J A. **1.189**
Lewis, H. **4.33**
Liang, Y C. **2.28**
Lieberman, A G. 55, **1.221, 2.10**
Liljeborg, A. **1.173**
Llebaria, A. **1.187**
Longuet-Higgins, M S. **3.3, 3.4**
Lou, D Y. **1.48**
Ludema, K C. **1.189**
Lukanowicz, C. **1.101**
Luo, N. xvi, **1.153, 2.14, 3.2, 3.19, 4.32**

McCool, J I. **1.143**
McCormick, N J. 29, **1.160**
McGivern, W H. **1.56**
McKuskey, R. **1.95**
Mc Mullan, D **1.18**
Madsen, L L. **1.205**
Mainsah, E. xv, **1.109, 2.7, 3.1, 3.2, 4.27**
Majumdar, A. **2.37**
Makosch, G. **1.75**
Mallick, S. **1.71**
Mandlebrot, B B. **2.34**
Manolakis, D G. **2.20**
Mantovani, J G. **4.13**
Marsden, G. **4.33**
Marsman, H J B. **1.175**
Martinez, A. **1.48**
Matey, J R. 55, **1.216**
Mathia, T. **2.7, 3.1, 3.2**
Matsuno, Y. **1.182**
Mattson, L. **1.90**
Maystre, D. **1.98**
Mich, L P. **1.146**
Mielczarek, S R. **1.194**
Mignot, J. **1.51, 1.126, 1.129**
Minsky, M. 5, 19, 126, **1.39, 4.30**

Mishima, S. **1.207**
Mitsui, K. **1.86**
Miyamoto, H. **1.207**
Miyamoto, K. **1.49, 1.52**
Moran, P J. **1.62, 1.66**
Morch, K A. **1.205**
Mulvey, T. **1.13**
Mummery, L. **1.150**
Musha, T. **1.49, 1.52**
Musselman, I H. **1.209**
Mygind, J. **1.205**
Myshkin, N K. **1.184**

Nakamura, M. **1.105**
Nankivell, J F. 43, **1.177, 1.178**
Nardin, P L. **1.129**
Nassirpour, F. **1.229, 3.9**
Nayak, P R. **2.24, 3.5**
Neumann, R. **1.206**
Nielsen, H S. **1.128**
Nomarski, G. **1.19**
Nowicki, B. **2.25**

O'Callaghan, P W. **1.132**
O-Hori, M. **1.185, 1.186**
Okada, T. **1.207**
Oostveldt, P V. **1.174**
Ozawat, N. **1.49, 1.52**

Palmari, J P. **1.187**
Pandit, S M. **1.229, 3.9**
Pantzer, D. **1.58, 1.59**
Passarelli, V. **1.162**
Patki, G S. **1.118**
Peklenik, J. 5, 10, 12, 125, **1.24, 4.29, 3.6, 3.7, 3.8, 4.29**
Perry, D M. **1.65, 1.66**
Peterson, P A. **1.209**
Peterson, R W. **1.62, 1.65**
Petran, M. 30, **1.176**
Pettigrew, R M. **1.55**
Peucker, K. **3.25**
Phillips, M J. **3.11**
Pickering, H W. **1.211**
Pohl, D W. **1.199**
Politch, J. **1.58**
Ponce, F A. **1.210**
Poroshin, V. **1.135, 2.2**
Prewett, A. **1.108**
Proakis, J G. **2.20**
Probert, S D. **1.22, 1.132**

Putman, C A J. **1.203**

Qian, B. **2.28**
Quate, C F. 126, 150, **1.83, 1.94, 1.225, 4.26**

Radcliffe, S J. **1.125, 1.131**
Radhakrishnan, V. **1.147**
Raja, J. **1.151, 2.32, 3.15**
Rasigni, F. **1.187**
Raski, J Z. **1.189**
Resandt, R W W V. **1.175**
Resnick, R. **1.221, 2.10**
Risko, D G. **1.221, 2.10**
Roberts, J I. **1.106**
Robinson, G M. **1.62, 1.65, 1.66**
Rohrer, H. 5, 47, 126, **1.80-82, 4.23**
Rondot, D. **1.129**
Ronnberg, A. **1.127**
Ronot, X. **1.166**
Roques-Carmes, C. **3.12, 4.4**
Rosenfeld, D P. **1.61**
Rondot, A C. **1.126**
Rose, J. **2.10**
Ruska. 41, **1.13**
Russell, P E. **1.209**

Sakurai, T. **1.211**
Sano, N. **1.211**
Santochi, M. **2.29, 2.30**
Sasajima, K. **1.121**
Sato, H. 45, **1.155, 1.156, 1.185, 1.186**
Savary, G. **2.36**
Sayles, R S. 5,10, **1.27, 1.116, 1.117, 3.10, 3.27**
Schneiker, C. **1.95**
Scire, F. **1.29, 1.124**
Scott, A C. **1.167**
Scott, P J. **3.26**
Shamir, J. **1.45**
Sherrington, I. **1.5, 1.47, 1.144, 1.153, 2.4, 2.5, 2.6, 2.17, 2.18, 2.21, 3.14**
Sherwood, K F. 54, **1.219**
Shimazu, H. **1.207**
Shotton, D. **1.161**
Shum, L Y. **1.220**
Silverman, P J. **1.213**
Simon, J. **1.41**
Simpson, J A. **1.42**
Smith, E H. **1.4, 1.5, 1.144, 1.154, 2.17, 2.18, 2.21, 3.14**

Snaith, B. **1.122**
Snelling, M A. **1.16**
Somekh, M G. **1.228**
Sommargren, G E. **1.57, 1.97**
Stanton, D. **1.48**
Stasiak, A. **4.12**
Stedman, M. **1.215**
Stelzer, E H K. **1.175**
Stengaard, I. **1.196**
Stewart, A D G. **1.16**
Stout K J. xv, **1.100, 1.109, 1.110, 1.111, 1.115, 1.119, 1.120, 1.133, 2.1, 2.7, 2.26, 2.31, 2.33, 3.1, 3.2, 3.13, 3.18, 3.19, 3.20, 3.29, 3.30, 4.1, 4.2, 4.5, 4.27, 4,32, 4.34**
Strecker, H. **1.195**
Stricker, R. **1.175**
Subarai, K. **2.7, 3.1**
Sullivan, P J. xvi, 153, **1.109, 1.110, 1.111, 1.115, 1.119, 1.120, 1.133, 1.135, 1.153, 2.1, 2.2, 2.7, 2.26, 2.31, 2.33, 3.1, 3.2, 3.13, 3.18, 3.19, 3.20, 3.29, 3.30, 4.1, 4.2, 4.7, 4.8, 4.27, 4.32**
Suzuki, A. **1.134, 2.15**
Swift, J A. **1.181**

Takase, T. **1.207**
Takenobu, T. **1.207**
Tantussi, G. **2.30**
Taylor, M L. **1.165**
Teague, E C. **1.3, 1.91, 1.98, 1.124, 1.194**
Tear, S P. **1.197**
Thomas, A P. **2.35**
Thomas, T R. 5, 10, **1.1, 1.27, 1.116, 1.117, 1.142, 2.22, 2.35, 3.10, 3.24, 3.27**
Thong, J T L. **1.183**
Thornton P R. **1.17**
Thwaite, E G. **1.99**
Tipton, H. **1.106**
Tolansky, S. **1.21, 1.53**
Tonder, K. **2.16**
Tramontana, J C. **1.210**
Travaglini G. **4.12**
Trumpold, H. **1.22**
Tsukada, T. **1.121**

Uchida, S. **1.155**

van de Leemput, L E C. **4.9**
van der Voort, H T M. **1.169**

Varnier, M. 1.187
Vazquez, L. 1.193
Vignale, M. 2.29
Virdee, M S. 1.157
Voelker, M. 1.95
Vorburger, T V. 1.3, 1.87, 1.91, 1.98,
    1.151, 1.221, 2.3, 2.10

Wadell, B C. 4.35
Wallach, J. 1.26
Ward, J. 1.29
Watanabe, M O. 1.202
Wehbi, D. 3.12, 4.4
Weibel, E R. 1.80, 1.81, 1.164
Weisenhorn, A L. 1.198
Werf, K O V D. 1.203
White, A D. 1.61
Whitehouse, D J. 1.92, 1.93, 1.141, 2.9,
    2.23, 3.11
Wickramasinghe, H K. 1.97, 1.223, 1.226
Wiechers, J. 4.11
Williams, C C. 1.223
Williamson, D E. 1.146
Williamson, J B P. 5, 10, 125, 1.23, 4.28
Wilson, T. 1.44, 1.50
Wintterlin, J. 4.11
Wirth, W M. 1.62
Wise, M L H. 4.5
Woods, A. 2.27
Wooley, R W. 2.12
Wu, S M. 1.229, 1.230, 3.9
Wyant, J C. 1.63, 1.64, 1.67, 1.69, 1.89,
    4.37

Xu, J. 1.224

Yakinci, M E. 4.36
Yamada, K. 1.182, 1.130, 4.3
Yanagi, K. 1.105
Young, R D. 5, 1.28, 1.29, 1.84, 1.85,
    1.91, 1.194

Zahouani, H. 3.2
Zhao, F L. 4.5
Zhou, G. 4.6